GLOBAL CHANGE RESEARCH NEEDS AND OPPORTUNITIES FOR 2022-2031

Committee to Advise the U.S. Global Change Research Program

Board on Atmospheric Sciences and Climate

Board on Environmental Change and Society

Division on Earth and Life Studies

Division on Behavioral and Social Sciences and Education

A Consensus Study Report of

The National Academies of
SCIENCES • ENGINEERING • MEDICINE

THE NATIONAL ACADEMIES PRESS
Washington, DC
www.nap.edu

THE NATIONAL ACADEMIES PRESS • 500 Fifth Street, NW • Washington, DC 20001

This activity was supported by contracts between the National Academy of Sciences and the National Aeronautics and Space Administration. Any opinions, findings, conclusions, or recommendations expressed in this publication do not necessarily reflect the views of any organization or agency that provided support for the project.

International Standard Book Number-13: 978-0-309-26134-0
International Standard Book Number-10: 0-309-26134-1
Digital Object Identifier: https://doi.org/10.17226/26055

Additional copies of this report are available from the National Academies Press, 500 Fifth Street, NW, Keck 360, Washington, DC 20001; (800) 624-6242 or (202) 334-3313; http://www.nap.edu.

Copyright 2021 by the National Academy of Sciences. All rights reserved.

Printed in the United States of America

Suggested citation: National Academies of Sciences, Engineering, and Medicine. 2021. *Global Change Research Needs and Opportunities for 2022-2031*. Washington, DC: The National Academies Press. https://doi.org/10.17226/26055.

The National Academies of
SCIENCES · ENGINEERING · MEDICINE

The **National Academy of Sciences** was established in 1863 by an Act of Congress, signed by President Lincoln, as a private, nongovernmental institution to advise the nation on issues related to science and technology. Members are elected by their peers for outstanding contributions to research. Dr. Marcia McNutt is president.

The **National Academy of Engineering** was established in 1964 under the charter of the National Academy of Sciences to bring the practices of engineering to advising the nation. Members are elected by their peers for extraordinary contributions to engineering. Dr. John L. Anderson is president.

The **National Academy of Medicine** (formerly the Institute of Medicine) was established in 1970 under the charter of the National Academy of Sciences to advise the nation on medical and health issues. Members are elected by their peers for distinguished contributions to medicine and health. Dr. Victor J. Dzau is president.

The three Academies work together as the **National Academies of Sciences, Engineering, and Medicine** to provide independent, objective analysis and advice to the nation and conduct other activities to solve complex problems and inform public policy decisions. The Academies also encourage education and research, recognize outstanding contributions to knowledge, and increase public understanding in matters of science, engineering, and medicine.

Learn more about the National Academies of Sciences, Engineering, and Medicine at **www.nationalacademies.org**.

The National Academies of
SCIENCES · ENGINEERING · MEDICINE

Consensus Study Reports published by the National Academies of Sciences, Engineering, and Medicine document the evidence-based consensus on the study's statement of task by an authoring committee of experts. Reports typically include findings, conclusions, and recommendations based on information gathered by the committee and the committee's deliberations. Each report has been subjected to a rigorous and independent peer-review process and it represents the position of the National Academies on the statement of task.

Proceedings published by the National Academies of Sciences, Engineering, and Medicine chronicle the presentations and discussions at a workshop, symposium, or other event convened by the National Academies. The statements and opinions contained in proceedings are those of the participants and are not endorsed by other participants, the planning committee, or the National Academies.

For information about other products and activities of the National Academies, please visit www.nationalacademies.org/about/whatwedo.

COMMITTEE TO ADVISE THE U.S. GLOBAL CHANGE RESEARCH PROGRAM

JERRY M. MELILLO (*Chair*, NAS), Marine Biological Laboratory
KRISTIE L. EBI (*Vice Chair*), University of Washington
ARRIETTA CHAKOS, Urban Resilience Strategies
PETER DASZAK (NAM), EcoHealth Alliance
THOMAS DIETZ, Michigan State University
PHILIP B. DUFFY, Woodwell Climate Research Center
BARUCH FISCHHOFF (NAS, NAM), Carnegie Mellon University
PAUL FLEMING, Microsoft
SHERRI W. GOODMAN, Woodrow Wilson International Center for Scholars, CNA
NANCY B. GRIMM (NAS), Arizona State University
HENRY D. JACOBY, Massachusetts Institute of Technology
LINDA O. MEARNS, National Center for Atmospheric Research
RICHARD H. MOSS, Princeton University
MARGO OGE, U.S. Environmental Protection Agency (ret.)
S. GEORGE H. PHILANDER (NAS), Princeton University
BENJAMIN L. PRESTON, RAND Corporation
PAUL A. SANDIFER, College of Charleston
HENRY G. SCHWARTZ, Jr. (NAE), Jacobs Engineering (ret.)
KATHLEEN SEGERSON, University of Connecticut
BRIAN L. ZUCKERMAN, Institute for Defense Analyses Science and Technology Policy Institute

National Academies of Sciences, Engineering, and Medicine Staff

AMANDA PURCELL, Senior Program Officer, Board on Atmospheric Sciences and Climate
AMANDA STAUDT, Senior Board Director, Board on Atmospheric Sciences and Climate
TOBY WARDEN, Director, Board on Environmental Change and Society
JENELL WALSH-THOMAS, Program Officer, Board on Environmental Change and Society
ALEX REICH, Associate Program Officer, Board on Atmospheric Sciences and Climate
RITA GASKINS, Administrative Coordinator, Board on Atmospheric Sciences and Climate
ROB GREENWAY, Program Associate, Board on Atmospheric Sciences and Climate

BOARD ON ATMOSPHERIC SCIENCES AND CLIMATE

MARY GLACKIN (*Chair*), The Weather Company, an IBM Business
CYNTHIA S. ATHERTON, Heising-Simons Foundation
CECILIA BITZ, University of Washington
JOHN C. CHIANG, University of California, Berkeley
BRADLEY R. COLMAN, The Climate Corporation
BART E. CROES, California Air Resources Board
ROBERT B. DUNBAR, Stanford University
EFI FOUFOULA-GEORGIOU (NAE), University of California, Irvine
PETER C. FRUMHOFF, Union of Concerned Scientists
VANDA GRUBIŠIĆ, National Center for Atmospheric Research
ROBERT KOPP, Rutgers, The State University of New Jersey
L. RUBY LEUNG (NAE), Pacific Northwest National Laboratory
JONATHAN MARTIN, University of Wisconsin-Madison
ALLISON STEINER, University of Michigan
DAVID W. TITLEY, U.S. Navy (ret.), Pennsylvania State University
DUANE E. WALISER, Jet Propulsion Laboratory

National Academies of Sciences, Engineering, and Medicine Staff

AMANDA STAUDT, Senior Board Director
LAUREN EVERETT, Senior Program Officer
LAURIE GELLER, Senior Program Officer
APRIL MELVIN, Senior Program Officer
AMANDA PURCELL, Senior Program Officer
ALEX REICH, Associate Program Officer
RACHEL SILVERN, Associate Program Officer
SHELLY FREELAND, Financial Business Partner
RITA GASKINS, Administrative Coordinator
ROB GREENWAY, Program Associate

BOARD ON ENVIRONMENTAL CHANGE AND SOCIETY

KRISTIE L. EBI (*Chair*), University of Washington
HALLIE C. EAKIN, Arizona State University
LORI M. HUNTER, University of Colorado Boulder
KATHARINE JACOBS, University of Arizona
MICHAEL ANTHONY MÉNDEZ, University of California, Irvine
RICHARD G. NEWELL, Resources for the Future
ASEEM PRAKASH, University of Washington
MAXINE L. SAVITZ (NAE), Honeywell, Inc. (ret.)
MICHAEL P. VANDENBERGH, Vanderbilt University
JALONNE WHITE-NEWSOME, Empowering a Green Environment and Economy, LLC
CATHY WHITLOCK (NAS), Montana State University
ROBYN S. WILSON, The Ohio State University

National Academies of Sciences, Engineering, and Medicine Staff

TOBY WARDEN, Director
ADAM JONES, Senior Program Assistant
TINA M. LATIMER, Program Coordinator
MARIA ORIA, Senior Program Officer
JENELL WALSH-THOMAS, Program Officer
JORDYN WHITE, Program Officer

Preface

This is a report from the National Academies of Sciences, Engineering, and Medicine Committee to Advise the U.S. Global Change Research Program (USGCRP) offering input to USGCRP on the development of its new 10-year strategic plan scheduled to be released in 2022. The report was prepared during "the year of COVID-19," 2020. As a result of this coincidence in timing, the committee spoke often about several parallels between the pandemic and the threat of climate change: It is global in scale, often hits the disadvantaged hardest, and requires scientifically informed and collective actions to avert the worst consequences. The themes of scale, equity, and science-to-action are woven throughout the report.

Over the past three decades, USGCRP has fostered coordinated research on all aspects of global change, especially climate change. The federal government has also supported U.S. engagement in collaborative international efforts of research, observation, and assessment. These efforts have resulted in impressive advances in understanding and robust modeling of global change and have also provided useful scientific knowledge to decision makers.

As impacts of climate change have become ever more apparent, the focus of USGCRP has evolved from a primary focus on the physical climate system toward the even more challenging focus on complex interactions among the physical climate system, Earth's ecosystems, and the human systems whose dynamics are governed by human actions. This continuing evolution supports USGCRP's mandate of assisting "the nation and the world to understand, assess, predict and respond to human-induced and natural processes of global change." A new USGCRP strategic plan that puts user needs at the forefront would entrain a broader and more diverse set of stakeholders and incentivize integrated research.

The Committee to Advise the U.S. Global Change Research Program is the body within the National Academies of Sciences, Engineering, and Medicine responsible for advising USGCRP. We are indebted to the staff at the National Academies who provided guidance, input, and support throughout the writing of the report, particularly Amanda Purcell, whose dedication and scientific understanding were critical throughout, and to Drs. Amanda Staudt and Toby Warden, whose deep technical knowledge and insights into the National Academies and USGCRP processes helped ensure an appropriately targeted report.

PREFACE

Finally, we dedicate this report to our late colleague, Dr. Anthony (Tony) Janetos, who chaired the Committee to Advise the U.S. Global Change Research Program from April 2017, until his passing too soon in August 2019. Over more than three decades, Tony wrote and spoke widely on the need to understand the scientific, economic, and policy linkages among the major global environmental issues and played an important role in shaping the public dialogue on climate change. We cherish the memory of Tony's thoughtfulness, his generosity of spirit, his insights, his humor, and his assuring smile.

Jerry M. Melillo, *Chair*
Kristie L. Ebi, *Vice Chair*
Committee to Advise the U.S. Global Change Research Program

Acknowledgments

This Consensus Study Report was reviewed in draft form by individuals chosen for their diverse perspectives and technical expertise. The purpose of this independent review is to provide candid and critical comments that will assist the National Academies of Sciences, Engineering, and Medicine in making each published report as sound as possible and to ensure that it meets the institutional standards for quality, objectivity, evidence, and responsiveness to the study charge. The review comments and draft manuscript remain confidential to protect the integrity of the deliberative process.

We thank the following individuals for their review of this report:

HALLIE C. EAKIN, Arizona State University
PETER C. FRUMHOFF, Union of Concerned Scientists
ISAAC M. HELD (NAS), Princeton University
JOHN P. HOLDREN (NAS, NAE), Harvard University
JEANINE A. JONES, California Department of Water Resources
THOMAS R. KARL, Climate and Weather, LLC
ROBERT KOPP, Rutgers University
MAUREEN LICHTVELD (NAM), University of Pittsburgh
FRIEDERIKE OTTO, University of Oxford
VARUN RAI, University of Texas-Austin
ROD SCHOONOVER, Ecological Futures
AMY K. SNOVER, University of Washington
JALONNE L. WHITE-NEWSOME, Empowering a Green Environment and Economy, LLC
KYLE WHYTE, University of Michigan

Although the reviewers listed above provided many constructive comments and suggestions, they were not asked to endorse the conclusions or recommendations of this report nor did they see the final draft before its release. The review of this report was overseen by **Katherine H. Freeman** (NAS), Pennsylvania State University, and **Dennis L. Hartmann,** University of Washington. They were responsible for making certain that an independent examination of this report was carried out in accordance with the standards of the National Academies and that all review comments were carefully considered. Responsibility for the final content rests entirely with the authoring committee and the National Academies.

Contents

Summary   **1**

1 Introduction   **15**
The Role of USGCRP in Preparing the Nation to Meet These Challenges, 17
A New Framework to Approach the Next USGCRP Strategic Plan, 18
Our Committee and Task, 21
Report Roadmap, 21

2 Global Change Risks to Human Systems   **23**
Population Health and Health Systems, 27
Food, 28
Water, 30
Energy, 31
Transportation and Infrastructure, 33
Economy, 34
National and International Security, 35
Integrating across Risks, 36
Implications of a Risk Framing of Research, 38

3 Integrated Systems-Based Research   **41**
Evolving USGCRP Priorities Toward an Integrated Systems-Based Approach, 42
Human System and Human-Natural System Science to Support Decision Making, 44
Designing and Implementing Integrated Systems-Based Research, 47

4 Research on Approaches Critical to Managing Climate Risk   **49**
Reducing Risk by Global Emissions Reduction, 50
Adaptation to Reduce Risks, 53
Solar Geoengineering Approaches, 55
A Need for Integrated Research on Risk-Management Approaches, 56

5 Crosscutting Research and Data Priorities   **57**
Extreme Events, Thresholds, and Tipping Points, 58
Simulation of Local- and Regional-Scale Climate, 63

Scenarios-Based Approaches, 66
Diversity, Equity, and Inclusion in Global Change Research, 70
Maintenance and Improvement of Data and Analysis Facilities, 71
Crosscutting Priorities to Advance Integrated Systems-Based
    Risk Management, 74

6 **Next Steps for Shifting the USGCRP Paradigm**   77
Organizational and Operational Changes, 79
Final Thoughts, 82

References   85

Appendixes
A   Statement of Task   97
B   Committee Member Biographies   99

Summary

Climate change is affecting the health and well-being of Americans across all parts of the country. Coastal areas are enduring more frequent and severe flooding due to sea level rise and storm surge; western states and Alaska have had increasingly devastating wildfires driven in part by hotter, drier, and longer fire seasons; and communities across the nation have suffered through extreme precipitation events and heat waves. These and other climate changes are posing risks to society—to people, their property, and their way of life—and to ecosystems, from croplands to national parks. In response to observed impacts and greater understanding of projected future challenges and opportunities, decision makers at local to national scales are considering how to reduce and manage societal risks associated with climate and other global changes in the coming decades by implementing a combination of mitigation and adaptation actions.

For more than three decades, the U.S. Global Change Research Program (USGCRP or "Program") has coordinated global change research across parts of the federal government. USGCRP, an interagency program established by Congress under the Global Change Research Act (GCRA) in 1990, consists of 13 federal agencies and departments and is overseen by the National Science and Technology Council. Fostered by USGCRP, interagency partnerships and collaborations with experts across the nation and the world have led to an unprecedented effort to observe, understand, predict, and project changes in natural and built environments.

In the fall of 2015, the National Academies of Sciences, Engineering, and Medicine's Committee to Advise the U.S. Global Change Research Program was asked to produce a report on the accomplishments of USGCRP over its first 25 years of existence. The committee highlighted four high-level examples of accomplishments in the report (NASEM, 2017a): supporting global observations systems; advancing Earth system modeling; increasing understanding of carbon-cycle science; and making progress toward the integration of the human dimensions of global change. The 2017 report also articulated the value-added of USGCRP research and coordination activities and the important evolution in their strategic planning to serve the needs of the nation.

The Program has continued to coordinate global change research activities across parts of the federal government, including establishing new interagency working groups on the water cycle and on understanding the dynamics of coastal systems. The most significant public accomplishment during the past 5 years was the release of the Fourth National Climate Assessment (NCA4) in 2017–2018.[1] The NCA4 provided many key updates on the state of the science and impacts of global change.

As mandated in the GCRA, the Program prepares decadal strategic plans laying out goals and priorities for federal research to advance scientific understanding and communicate information useful for policy decisions. The National Academies Committee to Advise the U.S. Global Change Research Program is charged to review the Program's draft strategic plans and to provide guidance to the Program on an ongoing basis. The committee prepared this report to inform USGCRP's thinking as it develops the next decadal plan, due to be completed in 2022.

Specifically, in preparing this report, the committee was tasked to consider how USGCRP can best meet the mandate[2] of the GCRA for the coming decade, in light of the significant climate change impacts happening today and the increases in their magnitudes and changes in their patterns that are projected over this time period, within the context of the longer-term changes projected in our climate and environment. This report identifies critical climate change risks, research needed to support decision making relevant to these risks, and opportunities for USGCRP's participating agencies and other partners to advance these research priorities (see Appendix A for the full Statement of Task).

The committee strongly supports ongoing efforts to observe, model, analyze, and communicate the physics and biogeochemistry of the climate system and the many mandated and other activities of USGCRP including conducting assessments. The committee assumes these efforts and activities will continue in the coming years.

In this report, the committee focuses on key risks the country could face in the 2030s, and highlights research needs that, if addressed, might increase resilience while supporting other societal goals, particularly the goal of reducing inequities. The committee has centered its advice in this report on how USGCRP could evolve to approach global change research differently in the coming decades, stressing that the largest risks expected will likely arise from the interactions of multiple systems, such as the food-energy-water nexus in the context of a changing climate. In addition, the report stresses that effective responses will arise from integration of social and natural sciences.

---

[1] NCA4 Part I, the *Climate Science Special Report* (CSSR), was released in 2017, and NCA4 Part II, the full assessment report that included the summary of the CSSR, was released in late 2018.

[2] To develop and coordinate "a comprehensive and integrated United States research program which will assist the Nation and the world to understand, assess, predict, and respond to human-induced and natural processes of global change." Section 101(b). Public Law 101-606(11/16/90) 104 Stat. 3096-3104.

Summary

This report does not provide a comprehensive list of global change research priorities, nor does it specify exactly how the Program should achieve the proposed evolution. The committee recognizes that USGCRP agencies and leadership are best positioned to identify how to accomplish this progression within the complex interagency environment, as proven by the Program's flexibility since its founding to evolve to meet the needs of the nation.

CLIMATE CHANGE POSES SIGNIFICANT RISKS TO AMERICAN SECURITY

Climate change currently poses risks to the American people, with projections indicating that each additional unit of warming will further increase these risks for nearly all impacts investigated. In this report, the term "risk" is used as defined by the Intergovernmental Panel on Climate Change (IPCC): "the potential for adverse consequences for human or ecological systems, recognizing the diversity of values and objectives associated with such systems"[3] (IPCC, 2019b, p. 696). Examples of risks in this report include those to health, food, water, energy, and transportation systems, and risks that affect the economy and national security. New research is needed to understand and communicate complex interactions among climate change (including uncertainties), other global changes such as disruption of the global nitrogen cycle, and societal development. Of special interest is the extent to which these interactions create immediate and urgent risks to Americans over the next decade, individually and collectively. Crucially, new research is needed on strategies to effectively and efficiently manage and reduce these risks in decades to come.

USGCRP is well positioned to help marshal the resources across multiple participating federal agencies, as well as other partners (e.g., state, local, and tribal policy makers), to support decision makers as they address these risks. Indeed, the Program has already taken steps in this direction, including past efforts to frame sections within the National Climate Assessments in terms of risk. That said, the committee believes that USGCRP and its participating agencies should make a significant pivot and center their next decadal plan, and the resulting priorities and activities, using an integrated risk-framing approach—that is, one that considers the risks to human and natural systems posed by climate change, and when appropriate, climate change together with other global changes. The committee also recommends focusing on and communicat-

---

[3] IPCC (2019b, p. 696) continues as follows: "Relevant adverse consequences include those on lives, livelihoods, health and well-being, economic, social and cultural assets and investments, infrastructure, services (including ecosystem services), ecosystems and species. In the context of climate change impacts, risks result from dynamic interactions between climate-related hazards with the exposure and vulnerability of the affected human or ecological system to the hazards."

ing the vulnerabilities and capacities of exposed systems and how these could shift over time, taking into account the multiple interconnections of projected changes, responses, and effects in human and natural systems. This approach is critical to effectively providing the information needed by decision makers at local to national scales.

> **RECOMMENDATION:** The committee recommends that USGCRP apply an integrated risk-framing approach to identify research priorities for the next 10 years that provide insights to avoid the worst potential consequences of urgent risks to human and natural systems from current and future climate change.

INTEGRATED SYSTEMS-BASED RESEARCH IS CRITICAL FOR MANAGEMENT OF CLIMATE RISKS

Decision makers in many levels of government, in private sector firms, and in society are increasingly requesting information on risks and responses to help them design and implement risk-reduction strategies. Traditional climate research that projects changes in the natural environment and then estimates the potential consequences of these changes for human systems, typically within sectors, is not fully meeting decision-maker needs. These projections rarely consider the complex multidirectional interactions among natural and human systems.

In this report, the term "natural systems" refers to the physical climate system and ecosystems (both unmanaged and managed, such as croplands), whose dynamics are governed by biological and/or physical processes. "Human systems" refers to systems managed by people to meet specific needs of society, and whose dynamics are governed by human actions. This report focuses on a set of human systems that evolved to meet specific societal needs: health, food, water, energy, transportation and infrastructure, the economy, and national security. These human systems interact with each other and with the physical climate system and ecosystems in complex ways through a series of drivers and feedback loops (see Figure S.1). The security of these human systems depends on their complex interdependencies, as well as interactions with natural systems. The management of risks to the coupled human-natural systems to increase their security for the benefit of society and the environment requires advances in scientific understanding of climate, ecosystem, and social and behavioral sciences.

Summary

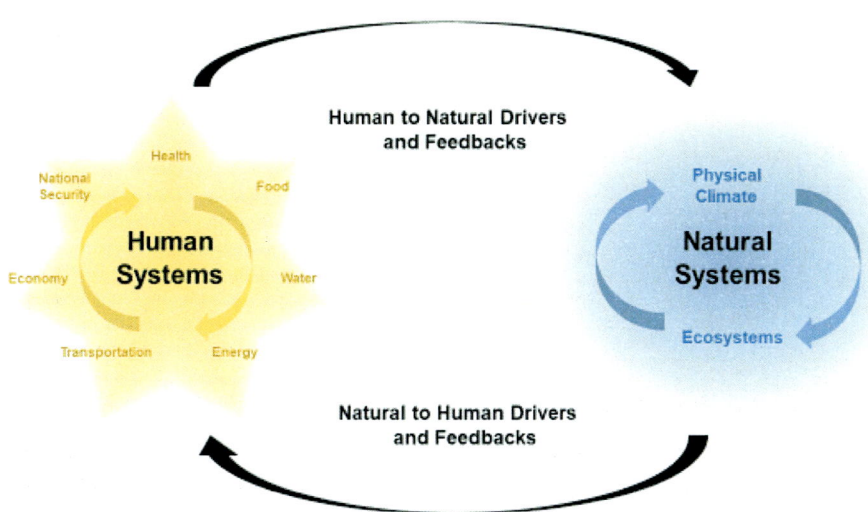

FIGURE S.1. Coupled human-natural systems are systems with interconnected, interdependent, and complex interactions among human systems, the physical climate system, and ecosystems. These interactions include the dynamics within one or more natural systems; the dynamics within one or more human systems; the processes through which the natural systems affect the human systems; and the processes through which the human systems affect the natural systems.

The committee identified seven critical human systems at particular risk from global change (discussed in detail in Chapter 2). "Systems" in this context refers to the individual human and natural systems, as well as the multidirectional coupling of human and natural systems; risks to these systems are the focus of Chapter 2. Effectively managing risks through mitigation and adaptation (focus of Chapter 4) would increase the security of these systems. Figure S.2 describes key terms used throughout the report.

Given its mandate to coordinate research across multiple agencies and the multiple dimensions of global change, USGCRP should play an important role in accelerating integrated, systems-based research. This research will be most efficient when coordinated with comparable international research and data collection efforts.

The Program has generated new insights since the last decadal plan (notably the sector-based assessments on food security [Brown et al., 2015] and impacts of climate change on human health [USGCRP, 2016]). However, meeting the urgent decision needs of the next decade will require increased commitment to research efforts that take a systems approach, involve collaborations among experts from across the health,

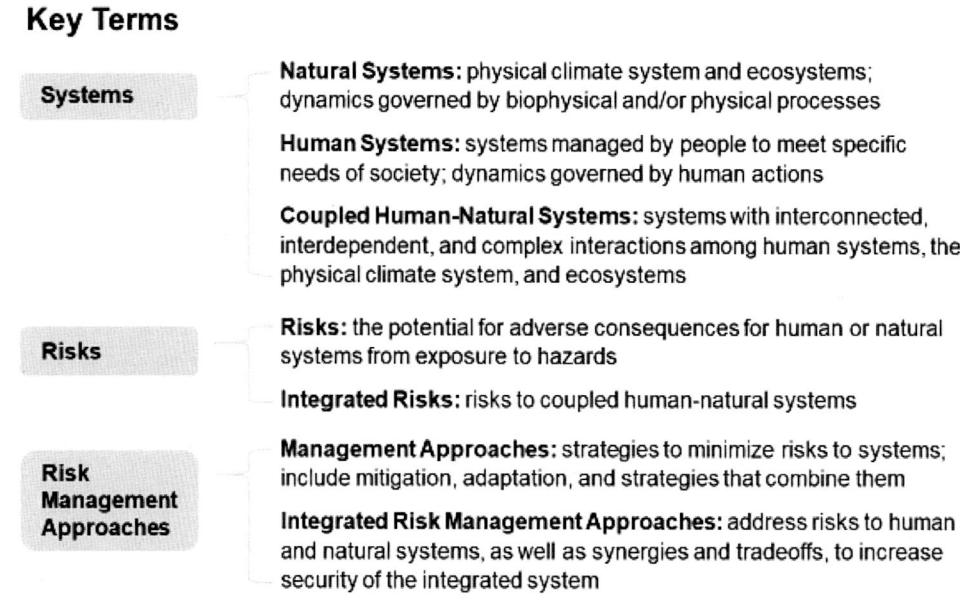

FIGURE S.2. This figure describes key terms used in this report: systems, risks, and management approaches for risks to systems. Each term can be applied at the individual and integrated level.

social, engineering, and natural sciences, and more explicitly consider the interactions among natural and human systems. Engagement with a range of stakeholders from local to national scales throughout the process will improve the usefulness and usability of insights generated.

The USGCRP mandate includes global environmental changes other than climate change; however, the primary focus of this document is on those risks posed by climate change. That said, the risk framing advocated here can be also applied to other global change issues and their interactions with social systems and climate change.

Driven by the urgency of addressing climate impacts happening today and projected risks for the near future, an integrated systems-based risk-management approach will enable USGCRP to more fully meet the mandate of the GCRA. This approach is the logical extension of the research priorities described in the Act and reflects the progression of knowledge and the advancement of data and research tools. Advances in fundamental and applied Earth system science over the next decade will be significantly more useful and usable by increased integration of natural and social sciences, improving the balance among physical climate research, ecosystems research, and human systems research.

> **RECOMMENDATION: The committee recommends that USGCRP accelerate the integration and communication of research on coupled human and natural systems to advance understanding of effective options for managing urgent climate change risks at local to international scales.**

New Research on Mitigation and Adaptation

Today, as climate consequences are clearly seen across the nation and the globe, there is an urgent need for policies to manage risks. New research is required to identify, communicate, and evaluate risk-management decisions covering the full range of potential policies that promote mitigation and adaptation strategies across local, national, and international scales. During the next decade, USGCRP will be increasingly called on to coordinate, rapidly advance, and communicate research in these areas, and to collaborate with global efforts.

Reducing Risks by Reductions in Global Greenhouse Gas Emissions and Lowering Their Atmospheric Concentration. Achieving net-zero emissions[4] of carbon dioxide from human activities is critical to managing climate change risks because this approach will inherently lower all future risks through avoided or captured emissions. Pursuing and informing mitigation-related policies will require better understanding of: (1) emissions targets that will avoid the most severe risks of climate change; (2) thresholds and tipping points in the climate system; (3) the socioeconomic risks of climate change, including thresholds and tipping points in social systems; (4) approaches for CO_2 removal, reliable sequestration, and utilization; (5) approaches to motivate effective uptake of policies and technologies; and (6) the ability to accurately quantify, and independently verify, the emissions of greenhouse gases (GHGs) at national and global scales. It is important to recognize that the strategies to reduce emissions and lower atmospheric concentrations also pose a range of associated risks to human-natural systems.

Increasing Resilience to Reduce Climate Change Risks to Americans. As stated in the GCRA, adaptation is an essential partner of mitigation in society's responses to global changes. Research on adaptation is critical for developing effective policies to manage the consequences of change. Effective policies at local to national scales require increased emphasis on access to, and effective uptake of, projected changes in climate and socioeconomic systems to inform decision support. Research and coordination

---

[4] Net-zero emissions are achieved when any CO_2 or other greenhouse gas emitted is offset by an equivalent amount of CO_2 removal and sequestration (NASEM, 2021a).

are needed to better understand: (1) the efficacy of adaptation practices implemented at local, state, federal, and tribal scales, and applied by industry and other actors; (2) what additional efforts are needed, today and in the future; (3) current and projected economic and social consequences of policy choices; (4) the processes of decision making to manage synergies and trade-offs over multiple scales; and (5) synergies and trade-offs between different adaptation and mitigation options.

In order to inform effective decision making that utilizes these strategies to reduce risk, increased understanding is needed not only on mitigation and adaptation options but also on the synergies and trade-offs between options. Additionally, decision makers might need to consider other strategies to reduce risk, such as solar geoengineering; advancing understanding to inform such decisions would likewise benefit from a highly integrated research strategy.

> **RECOMMENDATION: The committee recommends that USGCRP prioritize research related to managing climate risks, including (1) reducing global greenhouse gas emissions and lowering their atmospheric concentrations; (2) increasing resilience to current and anticipated climate-related security risks; and (3) expanding research on incentives for and the synergies and trade-offs between these risk-management approaches.**

CROSSCUTTING ANALYSES AND DATA ARE NEEDED TO SUPPORT MANAGEMENT OF CLIMATE RISKS

USGCRP has a long history of providing high-level coordination and communication of the research conducted at federal agencies through participating members in the Subcommittee on Global Change Research and interagency working groups established around priority focus areas for the Program. Given its role, mandate, and accomplishments in this coordination, USGCRP is especially well suited to make progress on additional crosscutting research efforts. These efforts would facilitate cross-comparison, provide consideration of the intersections of impacts (and responses) across multiple systems, and eliminate redundancy in underlying analyses.

An integrated, systems-based approach—that is, one that considers the multidirectional interactions among the physical climate system, ecosystems, and human systems—would benefit from the pursuit of several crosscutting priorities that can provide for the examination of challenges within and across the integrated systems. The report identifies five crosscutting areas that will contribute to addressing climate change risks: (1) extremes, thresholds, and tipping points; (2) simulation of regional- and local-scale climate; (3) a scenarios-based approach to project and manage climate change and associated risks; (4) equity and social justice; and (5) augmentation of existing analysis frameworks and supporting data.

Extremes, Thresholds, and Tipping Points. Extreme weather-related events have a range of societal impacts such as those associated with heatwaves, floods, storms, and wildfires. Research needs remain in projecting the frequency and severity of these events and improving attribution that links extreme events to natural and human-caused climate change. Of relevance to this research are the impacts of extreme events and tipping points on current and projected implications for a range of issues including asset values, human migration, conflict, and political instability.

Simulation of Regional- and Local-Scale Climate. Tools that provide the starting point for understanding of future climate—global climate models or Earth system models—have limited ability to simulate local-scale climate. However, societal harms from climate change very often occur on the local scale—for example, a low-lying area is flooded, or a town is destroyed by wildfire. There is an urgent need to improve capabilities to simulate local-scale climate, including hazards not represented, or not represented well, in global climate models such as coastal storm surge and wind intensity and direction driving wildfires. In addition, it is essential that these local-scale projections be presented in a manner that is useful and accessible to decision makers to increase resilience in their communities.

Refining a Scenarios-Based Approach to Project Climate Change, Associated Risks, and Effectiveness of Mitigation and Adaptation Policies at Global to Local Scales. Because managing climate change is an adaptive risk-management activity, it is important to project risks under a range of future climate and socioeconomic scenarios. Scenarios can be developed at local to global geographic scales and for decisions that will be taken at short to long timescales. Scenarios can be primarily model-based or can use models in participatory processes that include relevant communities to develop narratives and quantifications tailored to local decision-making needs. By championing state-of-the-art scenario-based models and processes that combine knowledge from across scientific disciplines and include a variety of relevant actors such as local stakeholders, USGCRP can help decision makers and others envision the risks associated with alternative combinations of GHG emissions and possible development pathways; these represent important uncertainties that need to be understood, quantified, communicated, and managed.

Equity and Social Justice. Climate change risk issues should be considered from the perspective of equity and social justice. Important issues include how extreme events and tipping points are and will be experienced differently across social groups, how mitigation and adaptation strategies have differential effects and might alleviate or exacerbate inequities, and how an equity and justice framing may increase the effectiveness of integrated risk management. The committee urges USGCRP to be attentive to multiple dimensions of equity and social justice, including race, ethnicity, indigenous status, gender identity, income and class, disability, age, and religion.

Augmenting Existing Analysis Frameworks and Supporting Data. Progress in research on global change risks, as well as the crosscutting topics, requires implementation of augmented analysis frameworks that can more adequately represent interactions among the physical climate system, ecosystems, and human systems. Enhanced data sets are needed to more adequately represent the many system interactions and provide results in forms that meet the needs of decision makers and the people they represent. More generally, there is a need for supporting and using advances in 21st century technologies, including big data management and related analytical technologies such as artificial intelligence, agent-based modeling (where appropriate), and emerging visualization approaches, as well as well-established approaches that are facilitated by emerging methods and technologies.

> **RECOMMENDATION: Expand research in five crosscutting areas: (1) extremes, thresholds, and tipping points; (2) regional- and local-scale climate projections; (3) scenario-based approaches; (4) equity and social justice; and (5) advanced data and analysis frameworks.**

SHIFTING THE USGCRP PARADIGM TO SUPPORT MANAGEMENT OF CLIMATE RISKS

The global crisis of the COVID-19 pandemic, its accompanying economic disruptions, and, in the United States, growing concerns with racial justice, inequality, and polarized politics demonstrate the need to envision and plan for multiple, often simultaneous, and multilevel disruptions to human systems, as well as to physical and ecological systems. Such preparation for multiple cascading risks requires interdisciplinary science more than ever, including the full range of disciplines across natural and social sciences. The ability for the nation to understand, adapt to, and respond to global changes will require investment from the U.S. research enterprise commensurate with the daunting challenges posed by the impacts of climate change on these interacting systems.

The committee recommends the Program employ an **integrated systems-based approach to risk management:**

- The **systems** on which the approach is based are coupled human-natural systems.

- The core focus of the committee's recommendations is managing **risks** of a changing climate to these systems, which are essential life-support systems for society.

- **Management** options for these risks include mitigation and adaptation and strategies that combine them.

- This risk-management approach should be comprehensive and **integrated**—considering benefits, trade-offs, path dependencies, and interactions among the risk-management components, their attendant uncertainties, and the interplay between and across the identified system. This needs to be done within a flexible framework that fosters the integration of the human and natural system components and the dynamic changes they will undergo through time.

This integrated systems-based approach is essential for understanding and communicating the complex consequences of concurrent mitigation and adaptation actions and their interaction. A comprehensive risk-management perspective will facilitate how the Program addresses emerging challenges posed by global change, including the co-benefits of mitigation actions and the synergistic and/or antagonistic results of multiple adaptation strategies, in ways that will be useful to and taken up by decision makers at multiple levels of society.

Meeting this ambitious update to its mandate will require a significant paradigm shift for USGCRP. Federal agencies that are already part of USGCRP will need to intensify their engagement in the Program, increasing involvement of suborganizations that bring relevant expertise and operational responsibilities to the table. It will also require greater participation of federal mission agencies that historically have not participated in USGCRP (e.g., U.S. Department of Homeland Security and its components such as the Federal Emergency Management Agency) but have relevant resources and expertise. In addition, it is critical that the next strategic plan outline the process through which participating agencies coordinate and adjust their individual program plans to avoid duplication and fill gaps critical to meeting overall program objectives. The committee recommends that the strategic plan should make clear the management structure and program criteria for setting priorities, sequencing investments, and guiding development of an integrated program across the individual agencies. This process should include input from user communities on a sustained basis consistent with effective engagement practices. This sustained engagement will require broader participation of user communities in the Program's planning and research that will effectively expand the coproduction approach already adopted by some USGCRP entities and will help identify research priorities based on the value of the information generated. This expansion of the next strategic research plan demands a rethinking of how the Program is organized and structured.

> **RECOMMENDATION: To accompany the shift in the USGCRP paradigm, the Program should explore organizational and operational changes to enhance the relevance and effectiveness of its work.**

Engaging local to national stakeholders throughout the cycle of setting research priorities is a core concept of this report. These stakeholders will assist in identifying effective implementation options and evaluating the degree to which evolving capacities will enhance the value of information generated. Throughout the process, special efforts are needed to counter inequitable distribution of risks, benefits, and costs across social groups.

Expansion of USGCRP would benefit from an analysis that considers a variety of approaches to growing the global change research enterprise to meet the evolving and intensifying challenges of global changes to society. Approaches to be explored in this analysis include: the reallocation of existing resources within the federal agencies and departments that make up USGCRP today; the inclusion of relevant federal agencies and departments not formally engaged with USGCRP; when warranted and possible, the acquisition of additional federal funds to support new research initiatives; and the fostering of public-private partnerships to expand the intellectual and financial resources supporting critical global change research.

> **RECOMMENDATION:** To enhance successful implementation of an integrated risk-management approach, it is critical that the Program does the following:
> 1. Prioritize diversity in both the Program and USGCRP activities by greatly expanding efforts to be inclusive and representative, and prioritize justice with research that highlights consequences and opportunities for underserved communities;
> 2. Increase the usability and relevance of research by adopting a coproduction approach to research, recommitting to the sustained assessment process, and establishing a standing user working group or advisory mechanism as a forum for input on user needs;
> 3. Advance program integration and accountability by increasing transparency of the management structure and criteria for setting priorities, sequencing investments, and guiding development of an integrated program across the individual agencies; and
> 4. Develop an evidence-based strategy for monitoring, evaluation, and learning for the Program's activities, including the next strategic plan, with flexibility for setting priorities and activities to adapt to and incorporate learning on an ongoing basis.

CHAPTER ONE

Introduction

The world's population is expected to increase by 2 billion people (about 25 percent), from 7.7 billion currently to 9.7 billion in 2050 (UN, 2019). Over the same period, the world economy is projected to more than double in size, far outstripping population growth (PwC, 2015). Already, a growing and wealthier human population is driving a set of interacting global changes that are disrupting the climate through: greenhouse gas (GHG)–emitting activities; polluting the air, land, and water; warming and acidifying Earth's oceans; and reshaping the land surface through cropping, forestry, and urbanization that threaten the planet's biodiversity.

These global environmental changes have already affected the health and well-being of the U.S. population, with many of these impacts driven by climate change. Residents of some coastal cities see their streets flood more regularly during storms and high tides and some see sea level rise as an existential threat. Inland cities near large rivers also experience more flooding, especially in the Midwest and Northeast. Insurance rates are rising in some vulnerable locations, and insurance is no longer available in others. Heat waves are the number one weather-related killer of Americans. Hotter and drier weather and earlier snow melt mean that wildfires in the West start earlier in the spring, last later into the fall, and burn more acreage. In Arctic Alaska, the summer sea ice that once protected the coasts is receding and autumn storms now cause more erosion, threatening communities with destruction and necessitating considerations of relocation.

Other global environmental changes, including global nitrogen pollution and land-use change, are already exacerbating risks to people and property, independent of and jointly with climate change. Increased nitrogen fertilizer application on croplands to increase yields not only has led to widespread surface and groundwater pollution but also has resulted in increases in the release of nitrous oxide—a potent GHG—along with carbon dioxide and methane, which together are major drivers of climate change. A dominant form of land-use change today is the clearing of tropical forests and woodlands for croplands. Much of the felled wood is burned on site, and the decay of soil organic matter accelerated. As part of this process, an estimated 1–2 billion metric tons of carbon are released to the atmosphere annually, amounting to 10 to 20 percent of human-caused carbon emissions (Global Carbon Project, 2020).

Projections of the magnitude and pattern of climate change published over the past decade paint a picture of a world increasingly challenged by extreme weather and climate events. An *extreme event* is a time and place in which weather, climate, or environmental conditions, such as temperature or precipitation, rank above a threshold value near the upper or lower ends of the range of historical measurements. Such events often have disproportional effects on people and the environment. The impacts of further climate change on food, water, health, and energy systems and on ecosystems are projected to be more extensive and severe than recent experience. At the same time, society's capacity to prepare for and manage these risks is increasingly challenged because of increases in the frequency and intensity of impacts, among other factors. Interactions between climate change and human systems (e.g., food, water, and health), and among human systems themselves, are complicated and interdependent. It is becoming ever more apparent that these complex interactions create cascading and compounding events and challenges that pose significant risks for Americans and people across the globe. Meeting these challenges requires a new framework for global change research—one that looks across and within natural and human systems to manage future risks.

Over the past decade, issues of equity and social justice have become increasingly important considerations in managing the risks of global change. The adverse impacts of global changes have been and will continue to be greatest on people who already suffer health and socioeconomic disparities. Moreover, climate change is now recognized alongside impaired air quality and hazardous waste as another driver of environmental injustice, racism, and disenfranchisement that undermines human well-being (Hoffman et al., 2020). As a consequence, climate risk and its change over time is not distributed equally within and among U.S. cities, regions, and economic sectors (Martinich and Crimmins, 2019). In addition, it is increasingly apparent that responses to climate change, including adaptation and GHG mitigation, can in themselves exacerbate social inequities and enhance vulnerability (Eriksen et al., 2021; Jakob et al., 2020).

The ability of the federal research enterprise to generate and communicate new insights that advance understanding of global change processes, the risks they pose to society, and the implications of alternative policies and technology is contingent on maintaining robust scientific infrastructure as well as the human capital that drives innovation. Scientific advances over the next decade can be enhanced by recognizing diversity, equity, and inclusion as being foundational to the U.S. Global Change Research Program's (USGCRP's or "Program's") mission. This will enable the nation to continue to attract world-class talent while integrating researchers with diverse expertise, experiences, and perspectives into the scientific enterprise.

THE ROLE OF USGCRP IN PREPARING THE NATION TO MEET THESE CHALLENGES

For more than three decades, USGCRP has been the home for coordinated research on all aspects of global change across the federal government. It also has supported U.S. engagement in collaborative international efforts of research, observation, and assessment. To spur the development of global observation systems, which are essential to research and analysis, USGCRP has fostered international cooperation through collaborations with organizations such as the Global Carbon Observing System. The collaborations have built and sustained unprecedented efforts to observe and document changes in the natural and built environments. Drawing on its extensive contacts in the domestic research enterprise, the Program plays a key role in marshaling participants from the U.S. scientific community to author and provide commentary for the Intergovernmental Panel on Climate Change and other international assessments. USGCRP's efforts, including its international contributions, not only have resulted in impressive advances in understanding and robust modeling of global change but also have brought useful scientific knowledge to bear in decision making (NASEM, 2017a).

As understanding of the processes and drivers shaping global change have evolved, so too has USGCRP. Strategic planning for the early years of the Program focused largely on major physical science questions driving global change and improving the ability to observe and model the changing climate. The Program's focus on "advancing science" has been essential for understanding the physical, chemical, and biological aspects of the Earth system; understanding how increasing atmospheric GHGs affects these systems; detecting human-caused changes in the observational record; and characterizing and quantifying the plausible uncertainties of the changes to come.

As impacts of climate change have played out over the globe, and across regions and sectors in the United States, USGCRP has conducted regular assessments of global change science and impacts that serve as a foundation of information for the nation's top decision makers, guiding future research investments and helping to prioritize policies to protect the health and well-being of Americans.

In 2015, the Committee to Advise the U.S. Global Change Research Program produced a report on the accomplishments of USGCRP over its first 25 years of existence. The committee highlighted the scientific advances discussed above, as well as articulated the value-add of USGCRP research and coordination activities and the important evolution in strategic planning to meet the needs of the nation. Since that report was released, the Program has continued to shepherd global change research activities across the federal government, including establishing new interagency working groups on the water cycle and on coastal systems and dynamics, and continued to coordinate U.S. participation in international efforts. The most significant public accomplishment during this time was the release of the Fourth National Climate Assessment (NCA4) in 2017–2018.

A NEW FRAMEWORK TO APPROACH THE NEXT USGCRP STRATEGIC PLAN

The Global Change Research Act of 1990 (GCRA; Public Law 101-606) established the Program and required that USGCRP produce a decadal strategic plan, with updates every 3 years. Relevant text from the GCRA (Section 104) is included in Box 1.1. At the time of the Act, an initial strategy was produced by the Committee on Earth Sciences under the George H.W. Bush administration (CES, 1989). The second decadal strategic plan was produced by the Program in 2003, with an update published in 2008. The third decadal strategic plan was published by USGCRP in 2012 to provide guidance for the period through 2021, an update to which was formally reviewed by this committee and published in 2016.

BOX 1.1
Excerpt from the Global Change Research Act of 1990

"SEC. 104. NATIONAL GLOBAL CHANGE RESEARCH PLAN.
(a) IN GENERAL—The Chairman of the Council, through the Committee, shall develop a National Global Change Research Plan for implementation of the Program. The Plan shall contain recommendations for national global change research…
(b) CONTENTS OF THE PLAN—The Plan shall—
1. Establish, for the 10-year period beginning in the year the Plan is submitted, the goals and priorities for Federal global change research which most effectively advance scientific understanding of global change and provide usable information on which to base policy decisions relating to global change;
2. Describe specific activities, including research activities, data collection and data analysis requirements, predictive modeling, participation in international research efforts, and information management, required to achieve such goals and priorities;
3. Identify and address, as appropriate, relevant programs and activities of the Federal agencies and departments represented on the Committee that contribute to the Program;
4. Set forth the role of each Federal agency and department in implementing the Plan;
5. Consider and utilize, as appropriate, reports and studies conducted by Federal agencies and departments, the National Research Council, or other entities;
6. Make recommendations for the coordination of the global change research activities of the United States with such activities of other nations and international organizations, including—

Introduction

> A. description of the extent and nature of necessary international cooperation;
> B. The development by the Committee, in consultation when appropriate with the National Space Council, of proposals for cooperation on major capital projects;
> C. Bilateral and multilateral proposals for improving worldwide access to scientific data and information; and
> D. Methods for improving participation in international global change research by developing nations; and
>
> 7. Estimate, to the extent practicable, Federal funding for global change research activities to be conducted under the Plan.
>
> (c) RESEARCH ELEMENTS—The Plan shall provide for, but not be limited to, the following research elements:
> 1. Global measurements, establishing worldwide observations necessary to understand the physical, chemical, and biological processes responsible for changes in the Earth system on all relevant spatial and time scales.
> 2. Documentation of global change, including the development of mechanisms for recording changes that will actually occur in the Earth system over the coming decades.
> 3. Studies of earlier changes in the Earth system, using evidence from the geological and fossil record.
> 4. Predictions, using quantitative models of the Earth system to identify and simulate global environmental processes and trends, and the regional implications of such processes and trends.
> 5. Focused research initiatives to understand the nature of and interaction among physical, chemical, biological, and social processes related to global change.
>
> (d) INFORMATION MANAGEMENT—The Plan shall provide recommendations for collaboration within the Federal Government and among nations to—
> 1. Establish, develop, and maintain information bases, including necessary management systems which will promote consistent, efficient, and compatible transfer and use of data;
> 2. Create globally accessible formats for data collected by various international sources; and
> 3. Combine and interpret data from various sources to produce information readily usable by policymakers attempting to formulate effective strategies for preventing, mitigating, and adapting to the effects of global change.
>
> (e) NATIONAL RESEARCH COUNCIL EVALUATION…
> (f) PUBLIC PARTICIPATION—In developing the Plan, the Committee shall consult with academic, State, industry, and environmental groups and representatives…".

The provisions in the Act outline what should be included as contents of the plan, list necessary research elements, and call for an evaluation of the plan by the National Academies of Sciences, Engineering, and Medicine (then the National Research Council). The Act also specifies that there should be public participation in the development of the plan, as well as information management recommendations, including guidelines for combining data from various sources "to produce information *readily usable* [emphasis added] by policymakers attempting to formulate effective strategies for preventing, mitigating, and adapting to the effects of global change."

USGCRP is approaching the end of the period provided for in its current decadal plan (2012–2021). As it begins planning for the next decade, laying the groundwork for decades to come, USGCRP has an opportunity to more proactively fulfill its mandate by producing an integrated science agenda essential for communicating information to enhance efforts at local to national scales as Americans work to manage the increasing and interacting challenges of further climate change. This re-orientation of USGCRP's strategy, which informs the structure of this report, is needed to provide enhanced decision support for those in the public and private sectors who are managing mounting risks of climate change on human and natural systems.

Traditional climate research that projects changes in the natural environment and estimates the potential consequences of these changes for human systems, typically within sectors, is not fully meeting decision-maker needs. These projections rarely consider the complex multidirectional interactions among natural and human systems. Integrated risk-based management research should be based on greater integration of the physical manifestations of climate change with ecological and socioeconomic changes. This research will be more effective if it focuses on the vulnerabilities and capacities of human and natural systems and how these will shift over time, taking into account the multiple interconnections of projected changes, responses, and impacts.

Ongoing input from users of the information on global change would ensure that a shift in the Program's orientation could help in identifying research priorities needed to meet its mandate to "provide usable information on which to base policy decisions relating to global change" (PL101-606). This information needs to be more than usable, it should be useful and easily used. In this report, the National Academies Committee to Advise the U.S. Global Change Research Program argues that these needs to manage climate change risks can be best met if USGCRP works within an integrated, systems-based risk framework and engages in coproduction with interested and impacted parties.

USGCRP has recognized the need for a risk-based approach to climate change challenges before. For example, the 2012 strategic plan aimed to "foster the iterative and collaborative dialogue between science and society needed to develop the scientific foundation for understanding and managing the risks of global change in the areas of greatest societal need" (USGCRP, 2012, p. 17). However, its risk-related focus was primarily on identifying and quantifying uncertainties within selected areas or sectors, rather than on understanding the widespread climate change risks discussed in this report, the connections among climate change risks, and effective and proactive means for addressing and managing them. A research program to meet these added challenges will require a more complete understanding of the multiple couplings among the physical climate system, ecosystems, and human systems than has been embodied in USGCRP research to date, one that recognizes the larger integrated system within which these risks arise and explicitly incorporates the needs and constraints of potential users of the research.

OUR COMMITTEE AND TASK

The National Academies have played a major role in shaping and advising USGCRP over the past three decades—from preparing scientific consensus reports in the 1970s and 1980s that led to the GCRA (which established the Program), to conducting regular document reviews of draft strategic plans and National Climate Assessments, to convening a standing committee forum for the Program leadership and participating agencies. These mechanisms have enabled the broader nonfederal scientific community's perspectives and expertise to be shared with USGCRP as it works to meet its mandate.

The Committee to Advise the U.S. Global Change Research Program provides ongoing and focused advice to USGCRP by convening key thought leaders and decision makers at semiannual meetings, providing strategic advice, and serving as a portal to relevant activities from across the National Academies. This committee is charged specifically to review draft strategic plans and updates thereof as requested; provide ongoing, integrated advice to USGCRP on broad, program-wide issues; and help to identify issues of importance for the global change research community.

To proactively meet its charge, this committee, with approval from the Program, developed this report to provide input in advance of the Program's next decadal strategic planning efforts (Statement of Task provided in Appendix A). The committee membership is broadly constituted to bring expertise in all of the areas addressed by USGCRP (see committee member biographies in Appendix B). To carry out its task, the committee has drawn on this expertise and the prior advice provided to the Program, as well as its reviews of Program assessments and other National Academies consensus reports.

REPORT ROADMAP

This first chapter highlights selected challenges for the United States in the coming decade arising from global change and offers an integrated approach for how USGCRP can prioritize investments into potential solutions for its next strategic planning period (2022–2031). Non-USGCRP agencies, corporations, and civil society can draw on the perspectives in this report to inform their own research agendas and investments. Chapter 2 outlines the most pressing global change risks in the coming decade identified by the committee and includes examples of integrating research needed to assess these risks. Chapter 3 discusses how USGCRP can use an integrated systems-based risk approach to provide more useful and usable information to help Americans deal with the urgent climate change challenge. Chapter 4 includes an overview of the portfolio of risk-management components to be considered within the mandate of USGCRP: mitigation techniques, adaptation needs for the set of previously identified human security risks, and examples of integrating needs to inform management of the climate change risks identified in Chapter 2. In Chapter 5, the committee provides five potential crosscutting priority areas for USGCRP to consider in addressing the nation's risk-management needs: extreme events and tipping points; improved simulation of local and regional-scale climate; a scenarios-based approach to project and manage climate change, associated risks, and effectiveness of mitigation policies; equity and social justice; and improved analysis frameworks and supporting data sets to meet these needs.

CHAPTER TWO

Global Change Risks to Human Systems

> **RECOMMENDATION: The committee recommends that USGCRP apply an integrated risk-framing approach to identify research priorities for the next 10 years that provide insights to avoid the worst potential consequences of urgent risks to human and natural systems from current and future climate change.**

The framing of global change has shifted from focusing on changes in land, oceans, atmosphere, polar regions, the planet's natural cycles, and deep Earth processes to understanding the risks created by interactions among the hazards created by these changes, the exposed regions and populations and their associated vulnerabilities, and the governance capacities to prepare for and manage changes in human and natural systems. "Natural systems" refers to the physical climate system and ecosystems (managed and natural). "Human systems" refers to systems managed by people to meet specific needs of society. This report focuses on a set of human systems that evolved to meet specific societal needs, including health, food, water, energy, transportation and infrastructure, the economy, and national security. These human systems interact with each other and with the physical climate system and ecosystems in complex ways through a series of drivers and feedback loops (see Figure 2.1). Risk is understood as *"the potential for adverse consequences for human or ecological systems, recognizing the diversity of values and objectives associated with such systems"*[1] (IPCC, 2019b, p. 696). A risk-management approach that is integrated would explicitly consider interactions across and among coupled human-natural systems, including benefits, trade-offs, and path dependencies.

The term "human security" was coined by the United Nations (UN) Development Program in 1994 as a conceptual framing to address multidimensional and complex soci-

---

[1] IPCC (2019b, p. 696) continues as follows: "Relevant adverse consequences include those on lives, livelihoods, health and well-being, economic, social and cultural assets and investments, infrastructure, services (including ecosystem services), ecosystems and species. In the context of climate change impacts, risks result from dynamic interactions between climate-related hazards with the exposure and vulnerability of the affected human or ecological system to the hazards."

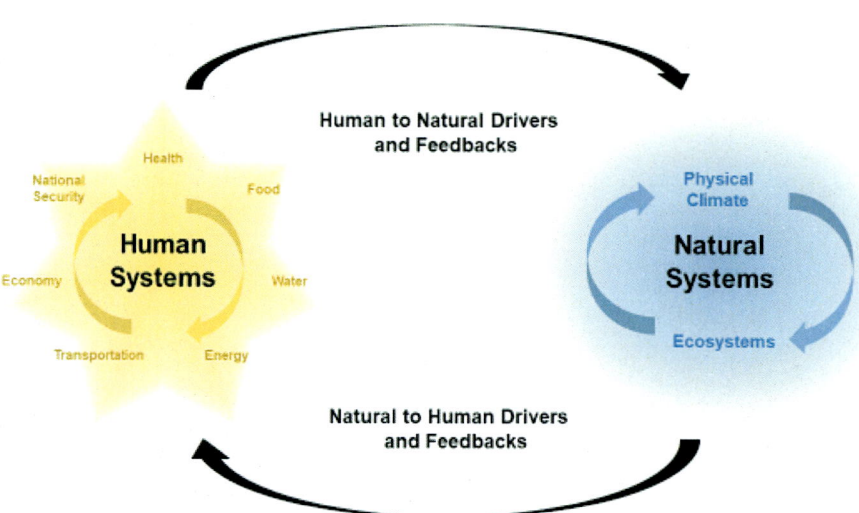

FIGURE 2.1. Coupled human-natural systems are systems with interconnected, interdependent, and complex interactions among human systems, the physical climate system, and ecosystems. These interactions include the dynamics within one or more natural systems; the dynamics within one or more human systems; the processes through which the natural systems affect the human systems; and the processes through which the human systems affect the natural systems.

etal challenges, such as climate change. In 2012, the UN General Assembly affirmed the value of human security as an approach to identify and address widespread and cross-cutting challenges to the survival, livelihood, and dignity of people.[2] More recently, it was applied as an operational tool for implementing the UN's 2030 Agenda for Sustainable Development (e.g., UN Trust Fund for Human Security, 2016). This human-centric approach encourages broad participation that provides detailed insights into the varying challenges populations face within communities and regions,[3] facilitating more targeted and community-driven solutions that address immediate vulnerabilities while building resilience and protecting livelihoods in the long term.

This chapter explores human security challenges through examples of global change societal risks to several human systems that will be important for the United States over the next decades. Throughout, the interdependence and interconnection of these risks, including through supply chains (across sectors), the particular vulnerabilities of frontline communities, and exposure to extreme events, are illuminated through a framework that calls for a greater coupling of the human and natural systems (see Fig-

---

[2] See https://www.un.org/humansecurity/wp-content/uploads/2018/04/What-is-Human-Security.pdf.
[3] See, for example, https://www.un.org/humansecurity/climate-change.

ure 2.1). Only by understanding this coupling can the greatest risks of global changes in the coming decade be effectively managed to reduce, to the extent possible, threats to the security of human-natural systems.

The U.S. Global Change Research Program (USGCRP or "Program") is well positioned to provide leadership in coordinating, integrating, and communicating research efforts across multiple sectors and agencies. The Program has taken promising steps recently to bring agencies together around three focal areas: water, coasts, and health. These efforts provide a foundation for the sort of integration that will be essential to addressing security risks and should be augmented with efforts to consider risks that cut across these focal areas, as well as the other risks identified here.

Throughout this chapter, the example of coastal communities is used to illustrate the ways in which the needs to understand risks are integrated across these human-natural systems (see Box 2.1 and other blue boxes throughout chapter). This is one of many possible examples and was selected to be illustrative rather than to imply that other integrated risks of are less concern.

BOX 2.1
Integrating Example: Security of Coastal Communities

Coastal communities house about 50 percent of the U.S. population at 3–4 times the population density of inland areas and have very large investments in built infrastructure as well as significant and fragile ecological resources, all of which are at risk from climate change (NOAA, 2013). Many coastal communities are already experiencing noticeable climate impacts (Sinay and Carter, 2020; Sweet et al., 2019) and more soon will (Neumann et al., 2015). Yet, despite the obvious climate-related dangers associated with coastal locations, continued population growth in these areas is expected (Aerts et al., 2014; NOAA, 2013).

Coastal communities and their leaders face a number of questions that a more integrated, multisectoral research approach could address (Sandifer and Scott, 2021). For example:

- How might recurring flooding affect transportation, housing, sanitation, energy systems, and health care, as well as work, recreational, and cultural opportunities and assets to the extent that a community is no longer perceived as livable?
- How might researchers design adaptive policy pathways that combine various low-regret, short-term actions that buy time (e.g., revising codes/standards, elevating structures, improving management of existing systems) with long-term adaptive solutions (e.g., building new infrastructure to hold back rising seas, relocating assets and populations)?
- How can lack of fairness in impacts and inequitable distribution of costs, benefits, and risks associated with responses be better understood and prevented?

Continued

> **BOX 2.1 Continued**
>
> An integrated approach to research can contribute to meeting these challenges, assemble and incorporate different kinds of knowledge and experience, go beyond technocratic solutions, and place more emphasis on pluralistic and comprehensive approaches to action-oriented knowledge for sustainability (Caniglia et al., 2021). In doing so, it will be essential to engage with affected communities to better understand local contexts and concerns and build trust in the analyses.
>
> Many research actions to assess risks from global change to the security of coastal communities are cross-sectoral and could be applied in various iterations to other communities. Examples of these needs are provided throughout this chapter in blue boxes.

Also crucial is the coupling of risks across nations. The USGCRP focus is on the United States, but what matters to the country is not limited to the direct effect of global change within the United States—for example, what the nation itself experiences through temperature and precipitation change, storms, increased disease, etc. Risks of global change for the United States are influenced by global change effects on other countries, how those countries are able to respond to them, and how risks are transmitted from across U.S. borders. Research needs to be not just global but also international in scope. For example, given the importance of international markets, food security is unavoidably a multinational challenge. Also, international research cooperation by the United States appropriately includes collaboration with research programs in other nations as well as aid for those nations lacking the resources to meet similar needs of their own societies. Opportunities for such contributions emerge in existing programs of international cooperation, and in assistance to developing countries for such activities as the preparation of national adaptation plans (UN LEG, 2019). Increasingly, the science needed for risk management by the United States needs to draw on research from across the globe; analyses within U.S. borders will not suffice.

A strong crosscutting message emerging from the committee's consideration of risk management was that risk management needs to focus on protecting the most vulnerable and reducing the underlying drivers of exposure and vulnerability, particularly inequities and exclusion. A variety of similarly interdependent and interconnected strategies are thus required to prepare for and manage these risks—for example, by direct risk reduction and increasing the resilience of these systems. These strategies need to be coordinated with and reinforce programs designed to directly address key vulnerabilities, such as tackling social injustices. These are discussed further in Chapter 4, but high-level, brief overviews of human-natural system security risks are provided here.

In its consideration of human-natural systems in the context of climate change, several USGCRP partners have explored the "nexus" framing approach, which is an integrative approach to systems planning and management that involves high complexity of scale, multiple stakeholders, and many processes. For example, the National Science Foundation developed a research program based on framing water, energy, and food as an interconnected system of systems in the face of climate change, as opposed to traditional silo-based resources planning and management approaches. A significant opportunity for USGCRP is to identify research issues that would benefit from understanding the interconnections and interdependencies involved in the complex and highly coupled systems and processes that affect society and the environment. Candidate examples of the nexus approach that integrate the climate system, ecosystems, and multiple human systems are briefly discussed in several of the sections in this chapter.

POPULATION HEALTH AND HEALTH SYSTEMS

Rising temperatures, changing precipitation patterns, increases in the frequency and intensity of extreme weather and climate events, sea level rise, and other global environmental changes are associated with increases in the numbers of cases of climate-sensitive injuries, diseases, and death. The Fourth National Climate Assessment concluded that the health and well-being of Americans are already affected by climate change. Key health risks include increased morbidity and mortality from heat waves and other extreme weather and climate events, adverse effects from exposure to poor air quality (including ozone and aeroallergens), effects on the emergence and distribution of vector-borne and other water- and food-borne infectious diseases, and consequences of reductions in the nutrient density of food and from undernutrition. There is also increasing recognition of how climate change is stressing mental health and well-being. Health risks are projected to increase with additional climate change (IPCC, 2018).

People and communities are differentially exposed to hazards and disproportionately affected by climate-related health risks. Populations experiencing greater health risks include children, older adults, low-income communities, and some communities of color (USGCRP, 2018). Climate change is exacerbating existing health disparities from social, economic, and environmental factors. Furthermore, many public health laboratories, health care facilities, and other infrastructure are at risk of damage and disruptions in service delivery during extreme weather and climate events.

Infectious disease emergence is an ever-present risk in a rapidly changing and interconnected world with increasing trade and travel, climate change, underfunded health systems, and urbanization (GHSI, 2019; Morand and Walther, 2020; Semenza et al., 2016). The emergence of COVID-19 is a dramatic example of this risk. There are hundreds of novel coronaviruses (Allen et al., 2017). Close human-wildlife interactions are key to the emergence of novel viruses into human populations, and that interaction is increasingly driven by demographic and global environmental change (Daszak et al., 2001; Loh et al., 2015). Recent outbreaks with significant impacts on health and economic security included SARS (2002), H1N1 influenza (2009), MERS-CoV (2012), H7N9 influenza (2013), Ebola (2014), Zika (2015-2016), and cholera in Haiti (2010–2019). Modeling future burdens of infectious diseases, particularly for new pathogens, is challenging because of the complexity of pathogen transmission (Ebi et al., 2018). Future risks for vector-borne diseases, such as malaria, dengue, and Lyme disease, could either increase or decrease with higher mean temperatures, depending on regional climate responses and disease ecology.

The health risks of a changing climate are current causes of preventable morbidity and mortality, which means health systems have policies and programs that could incorporate adaptation policies and programs to reduce the risks. Additional benefits to health arise from explicitly accounting for climate change risks in infrastructure planning and urban design (USGCRP, 2018).

> **Example Research Needs to Assess Global Health Security for Coastal Communities:**
>
> - Identify health care facilities and other health-critical infrastructure (e.g., potable water systems and wastewater treatment plants) that are vulnerable to disruption from flooding, storm surge, sea level rise, or electrical grid outages, and identify options to increase their resilience.
> - Develop approaches to study how information about observed impacts of extreme events and disasters, including the COVID-19 pandemic, can be used to improve locally applicable models and risk assessments for coastal communities.
> - Estimate the magnitude and pattern of health risks from extreme weather and climate events in coastal communities under a range of climate and development scenarios that include assumptions about transitions over the next decade and effectiveness of adaptation scenarios.

FOOD

By 2030, the impacts of climate change, other global environmental changes, and socioeconomic changes are projected to adversely affect food availability in the U.S. (IPCC, 2019a; USGCRP, 2017). Drivers of adverse changes include altering temperature and rainfall patterns, the frequency and intensity of climate extremes such as high temperatures and drought (see Section 4.1), and pest pressures. The magnitude and pattern of risks by mid-century will depend on the rate and severity of climate variability and change and on changes in trade, demographics, dietary preferences, and the extent to which effective mitigation and adaptation measures are implemented to address the growing challenges.

Risks to crop yields at 1.5°C above preindustrial temperatures could result in large transitions in land for food and feed crops and in pastureland, posing profound challenges for sustainable management of land for human settlements, food, livestock feed, fiber, bioenergy, carbon storage, biodiversity and other ecosystem services (IPCC, 2018). Risks at 1.5°C could be moderate to high for dryland water scarcity, soil erosion, vegetation loss, wildfire damage, tropical crop yield decline, and food supply instabilities (IPCC, 2019c). Increasing climate change also is expected to disrupt supply chains and negatively affect food production and prices, among other consequences. The extent of risk depends on socioeconomic choices as described in the Shared Socioeconomic Pathways (see Chapter 4 and O'Neill et al., 2016). Choosing plausible assumptions about international trade, demographic change, and food preferences is particularly important for projections of food security.

In addition, increased atmospheric CO_2 is reducing the nutritional quality of major cereal crops, including wheat and rice, reducing concentrations of protein, micronutrients, and B-vitamins (Loladze, 2014). At CO_2 concentrations expected later in the century, global projections indicated there will be hundreds of millions more people at risk of food insecurity and micronutrient deficiencies (e.g., Beach et al., 2019; Zhu et al., 2018).

Example Research Needs to Assess Food Security for Coastal Communities:

- Determine how climate change and extreme events will affect feedbacks among coastal social and ecological systems under a range of scenarios and project the implications on food webs and on local food production and supply.
- Determine the extent to which CO_2, climate change, and other global environmental changes could alter food security (not just crop yields) in the United States, in the context of potential changes in the global food system and possible domestic development choices. Coastal communities need estimates of the extent to which higher ocean temperatures and ocean acidification could affect fishing stock and seafood.
- Quantify the extent to which an increase in atmospheric CO_2 will continue to alter the nutritional quality of C3 plants,[a] and what this means for the health and well-being of coastal communities.

---

[a] C3 plants are plants in which the initial product of the assimilation of carbon dioxide through photosynthesis is 3-phosphoglycerate, which contains 3 carbon atoms.

Food crises are not just standalone problems; they also increase the risk of vector-borne and diarrheal diseases in children, put increased pressure on fragile terrestrial and aquatic ecosystems, and increase human-wildlife interactions that can drive the emergence of novel infectious diseases. Extensive agricultural trade means that decreased crop yields in one region can impact other parts of the globe. Food crises and famine also can result from and lead to political instability, migration, and conflict. Inequities within and between nations exacerbate these crises, creating groups especially vulnerable to disruptions in or inadequacy of the food supply.

WATER

Water security can be described as the ability of a population to maintain a reliable supply of clean water to sustain livelihood, well-being, agriculture, and ecosystems while adequately managing floods and droughts (USGCRP, 2018). Recent climate change has affected people's ability to sustainably access acceptable quality water during shortened or intensified rain seasons and their ability to protect themselves from water-related infectious diseases (Bakker, 2012; Thomas et al., 2013). Availability and access to water is becoming increasingly uncertain in regions as water stress is exacerbated by poor management of water resources and transboundary disputes. Water crises have already resulted in a lack of sanitation and increases in water-borne diseases,[4] food insecurity, conflict, financial instabilities (see, e.g., Gleick and Iceland, 2018), infrastructure damage, and biodiversity loss (see, e.g., UNEP, 2013). Most of these consequences will worsen with climate change (USGCRP, 2018). In 2018, drought was the second most costly hazard in the United States, with the greatest damages to the agriculture and livestock industry (NOAA, 2020). At the same time that climate change presents new challenges to water access and flood management, development and population growth are increasing demand for and vulnerability of water supplies. A lack of fresh water, both from precipitation and melting snowpack, will affect water storage, agriculture, wildlife, public health, and other critical factors.

Climate change affects the natural hydrological cycle through greater evaporation, the ability of a warmer atmosphere to hold more water, changes in atmospheric dynamics, reductions in seasonal snow cover, and more. These changes can result in increases in risk of flood and drought, sometimes both in the same location. In some places, shortened rain seasons will increase water demand and strain management capabilities. Where water is in abundance, locations vulnerable to flooding can experience saltwater intrusion, pollution, or destruction to infrastructure (CNA Corporation, 2017).

---

[4] See https://www.unwater.org/water-facts/water-sanitation-and-hygiene.

At the same time, warmer water in streams and rivers can impact the metabolism, life cycle, and behavior of aquatic species, in addition to causing disease, species loss, and increased competition from warm-water and/or invasive species. In the United States, warming global surface temperatures will lead to longer and more severe drought periods in the Southwest and other regions, and reduce spring snowpack in the mountains of the West (UCS, 2014).

Lack of access to clean water and sanitation occurs in the United States and is a major worldwide issue (Gasteyer et al., 2016). In the United States, there are particularly high rates of disparities in water access and sensitivity to climate impacts in BIPOC (Black, Indigenous, and people of color) communities. The water crisis in Flint, Michigan, is an example of this (see, e.g., Pauli, 2020 and Masten et al., 2016). In addition to water stress, other factors contributing to livelihood activities will be adversely affected. For example, water has been weaponized in situations of conflict to pursue security interests. Boko Haram has poisoned water sources, ISIS has controlled dams in water-scarce areas, and drug traffickers in Guatemala blocked parts of rivers for transport of contraband (CNA Corporation, 2017).

With each degree of warming, it is estimated that renewable water resources will decrease by about 20 percent for an additional 7 percent of the world population (IPCC, 2014). As water issues persist, they will negatively influence public health and drive economic, political, and social instability. And, as with other climate-driven risks, water insecurity is exacerbated by multiple dimensions of inequality.

> **Example Research Needs to Assess Water Security for Coastal Communities:**
> - Identify and quantify ways in which climate change can impact water use and management by governments and the role of ports and coastal communities in the global supply chain.
> - Refine models to project how extreme precipitation events and hot weather will contribute to overwhelming sanitary sewers, toxic algal blooms, inundation, and challenges to water safety and security for coastal communities.

ENERGY

The energy sector is undergoing rapid change including fuel switching from coal to natural gas, electrification of the vehicle sector, increased deployment of renewable energy, increased energy efficiency in most sectors, changes to the electric grid, and changes in response to the dynamics of international energy markets (NASEM, 2021a). In addition, the emergence of COVID-19 significantly altered expectations regarding future trajectories of U.S. and global energy demand (IEA, 2021). Declines in energy use (both electricity and transportation fuels) due to changes in consumer behavior could have long-term impacts. Furthermore, the global economic shock induced by COVID-19 constrained opportunities for investment in the sector over the near term but also disrupted assumptions regarding the sustainability of some energy portfolios over the long term (Hepburn et al., 2020; Hosseini, 2020; Zhong et al., 2020). For example, the adverse economic impacts of the pandemic were largely borne by coal, oil, and natural gas producers and fossil fuel electricity generators (IEA, 2021). Meanwhile, demand for electricity from renewables increased.

The rapidly changing energy sector is already interacting with the changing climate to create opportunities and challenges. Extreme weather conditions represent the most common source of electricity outages in the United States, including severe winter storms, tropical storms and hurricanes, heatwaves, and wildfires (DOE, 2017). For example, flooding from Hurricane Harvey forced oil refineries along the Texas Gulf Coast to shut down temporarily in 2017. That same year, Hurricanes Maria and Irma caused catastrophic damage to the electricity grids of Puerto Rico and the U.S. Virgin Islands (Campbell et al., 2017; Clarke et al., 2018). Recent catastrophic wildfires in California during 2017, 2018, and 2019 were attributed in part to extreme weather and electricity infrastructure failures. Such extreme events are projected to grow in intensity, frequency, and/or duration as the climate changes (USGCRP, 2017). In addition, extreme events interact with more chronic pressures such as sea level rise that have the potential to increase the risk of temporary or permanent inundation of coastal energy infrastructure (DOE, 2014; Government Accountability Office, 2014; Maloney and Preston, 2014).

Example Research Needs to Assess Energy Security for Coastal Communities:

- Enhance climate and weather prediction and projections for the energy sector over different timescales in response to plausible scenarios of energy demand, energy technologies, greenhouse gas emissions, land-use change, and demographic and technology change (e.g., energy needs for cloud storage and new electronic devices).
- Improve understanding of energy innovation, socioeconomic trends, and interdependencies among energy and other critical infrastructure systems and their implications for energy security, vulnerability, and risk management.
- Improve understanding of opportunities and constraints associated with the retirements of aging energy assets in coastal communities as well as the siting of new energy technologies to enhance resilience and reliability.

TRANSPORTATION AND INFRASTRUCTURE

Protection of the built infrastructure (i.e., transportation, energy, water, wastewater, and communication systems) is critical to the continued security of the nation's economy and social fabric. Of particular concern are the transportation networks—the roads, highways, bridges, ports and harbors, airports, rail, and pipelines that form a system of systems at local, regional, and national levels. Climate change, especially sea level rise and extreme weather events, will very likely have increased impacts on the country's transportation systems.

Disruption of the transportation system, or parts thereof, can affect virtually all aspects of peoples' lives—from access to and utilization of health care, to food and medicine distribution, to industrial supply chains, to emergency evacuation. Sea level rise places coastal transportation systems, as well as communities and businesses, at increased risk, while extreme events, such as floods and fires, can shut down the transportation systems in most areas of the country, sometimes for weeks or months. Often the most vulnerable segments of the population experience the greatest proportion of the impacts, exacerbating the risks they already face.

Development of effective policies and solutions for the threats of climate change to infrastructure is a complex task (NASEM, 2016c). Among the principal considerations are the interactions and interdependencies among transportation, energy, the built infrastructure, the economy, and other human systems. Uncertainty in climate science and the interactions of natural and human systems require a different approach to decision making (discussed further in Chapter 5). Additionally, there is a fundamental need for sound risk-based asset management to evaluate the vulnerability of a given system or subsystem, the likelihood of disruptive events, the consequence of the disruption, and the cost and means to reduce impacts. Situations will vary from high probability, low consequence events to low probability, high consequence events (DOT and FHWA, 2013). For example, what are the critical nodes, damage to which will result in major or cascading impacts, not only to the system itself but to interrelated physical and human systems, and in particular to vulnerable populations? Research is needed to identify more resilient materials and/or systems that can better withstand climate changes, including redundant systems and for critical nodes. Indeed, a basic requirement is determining a "sufficient" level of resilience for systems and subsystem components. The committee also notes that because a substantial part of infrastructure investment depends on local or regional resources, considerable and consequential inequities in ability to respond to these risks can arise.

ECONOMY

Human well-being ultimately depends on, among other things, access to basic human needs, such as food, water, housing, health care, and communities. Direct access to many of these essential components of economic security often requires not only that they be available (i.e., supplied), but also that those in need have the income or other resources necessary to purchase goods and services when provided by markets. This highlights the importance of economic security as an additional security concern.

Economic security is often defined as "the degree to which individuals are protected against hardship-causing economic losses" (Hacker et al., 2014, p. S7). When individuals experience income disruptions, they can suffer hardships or losses that have significant impacts on their well-being, often well beyond the associated financial losses (e.g., Helliwell et al., 2014). Moreover, those impacts can have ripple effects throughout communities and economies, as income losses within one group spill over to other groups. Economic insecurity has been generally rising since the 1980s (Hacker et al., 2014), but with some subgroups much more vulnerable, such as communities disadvantaged and/or marginalized because of race or ethnicity (Chetty et al., 2018; McIntosh et al., 2020). The recent COVID-19 pandemic has highlighted how shocks can cascade when sectors, regions, and countries are closely interdependent through demand and supply, and when stay-at-home mandates fundamental for health protection significantly interrupt economic activity.

Climate change and associated extreme events can contribute to economic insecurity through supply disruptions at the production, distribution, or consumer levels. For example, increased frequency and severity of droughts and flooding create greater risk of crop losses, which threaten not only food availability but also the livelihoods of agricultural producers and workers. Likewise, extreme events such as hurricanes can create major income shocks to those directly affected (e.g., through losses of uninsured assets) and to those indirectly affected (e.g., through job or business losses [Bleemer and van der Klaauw, 2019]).

These losses can last for years after an event. For example, even 10 years after Hurricane Katrina, residents of New Orleans whose property was flooded had higher rates of insolvency and lower homeownership than their non-flooded neighbors (Bleemer and van der Klaauw, 2019). Hurricanes can also significantly affect migration, which can in turn have economic implications in both the origin and destination areas (Fan and Davlasheridze, 2018) and ultimately determine long-run economic impacts on affected individuals (Deryugina et al., 2018).

> **Example Research Needs to Assess Economic Security for Coastal Communities:**
>
> - Increase research on the socio-political impacts of economic insecurity in coastal communities, particularly the potential for associated large losses of jobs and/or income that could trigger social unrest and conflicts (Harari and Ferrara, 2018).
> - Enhance understanding of both supply-related disruptions and disruptions due to lost wage income or large expenditure shocks (e.g., medical, housing or moving costs) in more vulnerable subpopulations of coastal communities.

NATIONAL AND INTERNATIONAL SECURITY

By 2030, without the employment of rapid mitigation or carbon-removal strategies, climate change is projected to, directly and indirectly, affect the national security environment, its institutions, and infrastructure (IMCCS, 2020). Climate change is a diffuse threat that cannot be addressed by engaging with a single actor. The recognition of the crosscutting risks climate change presents to the intelligence and military community elucidates how a new national security paradigm—of which climate change is a bedrock component—is evolving (Werrell and Femia, 2019). The impacts of the COVID-19 pandemic offer an example of how a nonmilitary threat like infectious disease emergence, which is affected by climatic conditions (WHO, 2003), can cause global social disruption, insecurity, and recession (Klarevas and Clarke, 2020). Similarly, emerging studies on how environmental destruction and disruption, worsened by climate change, can contribute to recognized security threats, has widened the security challenge (Coats, 2019).

Climate change is also a threat multiplier that can negatively influence existing risks to U.S. and global security (CNA Corporation, 2007). The intensification of water scarcity, temperature rise, precipitation inundation, wildfires, and other climate-related events can further aggravate emerging state instability and failure, interstate tension, conflict, military intervention, and other high-order security risks if not addressed (Guy et al., 2020). A changing environment can destabilize a region and influence local resource competition, land degradation, food and water availability, livelihood instability, and more by altering global human activities. Citizen dissatisfaction, mismanagement of resources, government destabilization, and violence are then compounded by climate change's regional impacts, increasing the likelihood of conflict (Center for Climate and Security, 2019).

Escalated tensions may ensue over territorial claims between regional powers fighting to gain control of natural resources needed to sustain local livelihoods (UN Interagency Framework Team for Preventive Action, 2012). These climate-related impacts add to existing tensions as populations are stripped of their basic needs, bringing about heightened regional and global threats. Climate change increasingly contributes to state fragility that can then be taken advantage of by terrorists, insurgents, and/or transnational criminal groups (Nett and Rüttinger, 2016). Security risks present in one region then can spill over into nearby fragile areas through humanitarian demands and population movement.

Challenges to both military and civilian infrastructure caused by climate-related risks stem from the disruption or destruction of physical and network infrastructure that can unsettle or weaken regional security (IMCCS, 2020). In low-lying regions of the world, sea level rise and storm surge threaten infrastructure that serves millions (UCS, 2016). Damage to ports and military installations can negatively influence trade and military preparedness as supply chains are disrupted and troops are left unable to deploy (Guy et al., 2020). Similarly, during natural disasters, disruptions in health care, transportation, communication, water treatment, energy, and other infrastructure will exacerbate existing and emerging health issues (NIC, 2017). Military headquarters, logistic hubs, and joint task forces that face extreme climatic stress will exhaust their abilities to carry out operations (Fetzek and Schaik, 2018). If unmitigated, climate change risks can adversely affect international, national, and, more generally, human security.

> **Example Research Needs to Assess National and International Security for Coastal Communities:**
>
> - Understand the robustness of local security institutions and infrastructure to further stresses in order to identify when tipping points could be crossed and their possible consequences as the numbers of conflicts and crises requiring response multiply.
> - Identify which nations are likely most susceptible to climate changes, including sea level rise, considering the intersection of climate change, policy responses of other nations, nuclear proliferation, potential adaptation, and others.

INTEGRATING ACROSS RISKS

As evident in the previous sections, each of these security risks has multiple intersections with other domestic risks and with similar impacts in other countries. These compound risks arise from global to local interactions resulting from the worldwide exchange of people, goods, money, information, and ideas across human and natural systems including infrastructure (physical and digital), financial institutions, natural resources, manufacturing, food production, biodiversity, climate, and other systems

(World Economic Forum, 2020). The complexity of the interactions and underlying systems creates interdependencies that "we do not understand and cannot control well" (Helbing, 2013, p. 51). Research programs need to be designed to model and understand how one system propagates risks to other interconnected systems (Haimes, 2018). Information on how systems are interconnected will better inform decisions at these intersections. Integration requires shifting the focus to the vulnerabilities and capacities of single systems or sectors to interconnected systems and how these will shift over time, taking into account the multidirectional interactions of projected changes, responses, and effects. High resolution multimodel frameworks and analysis tools are needed to understand how human and natural systems co-evolve in response to environmental, technological, and societal transitions and shocks and what approaches can manage the resulting interdependent risks across sectors and geographies. USGCRP (2016) explores the potential for interagency collaboration focused on developing a conceptual framework to integrate models and empirical studies of interdependent systems and the various levels of detail, complexity, and spatiotemporal resolution needed to address specific risk-management approaches.

As risks rise, decision makers will increasingly need to manage and communicate synergies and trade-offs between policies that are potentially beneficial for one sector but harmful in another. Blue boxes throughout the preceding sections provide examples of research questions in each of the security risk areas. These examples illustrate the sort of multidisciplinary research accorded by a risk-framing approach, as well as interconnections among risk areas.

For example, numerous factors across human and natural systems individually and collectively affect the vulnerability of coastal communities to climate change (see Box 2.1), including (1) elevation; (2) rate of locally apparent sea level rise; (3) history of and likely future exposure to extreme weather events, saltwater intrusion, harmful algal blooms, infectious disease organisms, oil spills, chemical contamination, and ocean acidification; (4) susceptibility of critical life- and health-sustaining infrastructure; (5) porous soils and subsidence; (6) history of socioeconomic deprivation; (7) availability and cost of property insurance and financing; (8) political decision making, policies, and regulatory structures of federal, state, and local agencies in relation to coastal development and protection; and (9) effectiveness of community leaders. Many of these factors are also relevant for human health and food and water security; thus, steps to assess, advance understanding of, and manage coastal risks also need to consider implications for these other risk areas (see blue boxes in this chapter).

USGCRP is well positioned to provide leadership in coordinating and integrating research efforts across multiple sectors and agencies. The Program has taken promising steps recently to bring agencies together around three focal areas: water, coasts, and health. These efforts provide a foundation for the sort of integration that will be essential to addressing security risks and should be augmented with efforts to consider risks that cut across these focal areas, as well as the other risks identified here. But many challenges remain for coordinating and integrating research efforts across multiple sectors and agencies. For example, efforts to take account of justice and equity and to engage with decision makers and stakeholders need to be expanded.

IMPLICATIONS OF A RISK FRAMING OF RESEARCH

USGCRP is mandated to help marshal and coordinate resources across multiple participating agencies, in cooperation with similar efforts in other nations, to address risks. Indeed, the Program has already taken some steps in this direction, including efforts to frame sections within the National Climate Assessments in terms of risk. Communicating effective approaches to managing climate-related risks, and ultimately reducing these risks in decades to come, will require robust information and understanding of the physical climate system, ecosystems, and human systems. While USGCRP does not directly manage risks, in setting its research priorities, USGCRP can and should seek to identify information that, when communicated, would be most useful, usable, and impactful at local to national scales. Adopting a broadly defined "value of information" perspective can help not only to maintain a focus on the use and usefulness of the information and insights gained through research, but also to ensure that scarce research resources are allocated so as to be most beneficial in managing risks (e.g., Cooke et al., 2014; Keisler et al., 2014; Rushing et al., 2020). The committee emphasizes again that linking research and research planning to deliberation with interested and impacted parties has proven an effective approach.

> **Examples of Integrating Needs to Assess Global Change Risks to Coastal Communities:**
>
> - Understand how risks propagate across human and natural systems, and the levels of detail, complexity, and spatiotemporal resolution needed to model and manage risks.
> - Develop approaches to study how information about observed impacts of extreme events, including the COVID-19 pandemic, can be used to improve locally applicable models and risk assessments for coastal communities.

Using a risk framing for strategic planning and priority setting, as well as in its assessment activities, USGCRP would prepare the nation for urgent and immediate risks, as well as those projected to occur over the medium term. Centering the next decadal research plan on integrated risk management will require USGCRP and its participating agencies to achieve better alignment and coordination.

The following two chapters discuss these two implications and provide suggestions about how USGCRP can put a risk framing into practice. In the committee's view, adopting an integrated systems-based risk framing could help ensure that the research outputs of participating USGCRP agencies provide the necessarily integrated information that supports the preparation of assessments, facilitates the synthesis of research and other sources of information to support decision makers at all levels of governance (including those in the private sector), and develops a specific climate change risk-reduction framework.[5] These efforts will be key to ensuring that decision makers have the information they need to manage global change risks in an integrated fashion across time scales, taking into account the synergies and trade-offs among the challenges to systems.

---

[5] The United Nations' Sendai Framework for Disaster Risk Reduction (UNDRR, 2021) provides a model for national- and global-level disaster risk reduction that may be useful in informing a climate change impact risk-reduction framework.

CHAPTER THREE

Integrated Systems-Based Research

> **RECOMMENDATION: The committee recommends that USGCRP accelerate the integration and communication of research on coupled human and natural systems to advance understanding of effective options for managing urgent climate change risks at local to international scales.**

Decision makers in many levels of government, in private sector firms, and in society are increasingly requesting information on natural and human systems and their multiple interconnections to help them design and implement risk-reduction strategies. Typical approaches to climate research that project changes in the natural environment and then estimate the potential consequences of these changes for human systems, typically within sectors, are not meeting their needs (Holm and Winiwarter, 2017). These projections generally do not consider the complex coupling between natural and human systems and often do not consider how societal systems are likely to evolve over coming decades. Furthermore, differential impacts across groups and thus equity impacts of climate change and responses to climate change (mitigation and adaptation) are often ignored.

Effectively managing climate risks requires greater integration of the physical manifestations of climate change with its ecological and socioeconomic consequences. Factors to be considered include vulnerabilities and capacities of exposed systems; multidirectional coupling interactions; multiple interconnections of projected changes, responses, and effects in human and natural systems; and the implications of these dynamics for equity and social justice.

This chapter highlights research in human and natural systems, as well as their interactions, that are underemphasized in the current U.S. Global Change Research Program (USGCRP or "Program") activities and that could put the nation at risk over the coming decade in the absence of further research and investment. Building new capabilities over the next decade will not only increase resilience over the time period covered in the next strategic plan but will lay the foundation for resilience to risks that are projected to arise or increase by mid-century. The committee recognizes that the current activities agencies formally consider part of the USGCRP portfolio are not well aligned with these areas and therefore identifies ways the Program can accelerate its transition in this direction. Given the urgency of climate risks, the nation can no longer afford for the Program's historical interpretation of the U.S. Global Change Research Act (GCRA) to constrain efforts to build the capabilities needed to provide useful and usable information.

EVOLVING USGCRP PRIORITIES TOWARD AN INTEGRATED SYSTEMS-BASED APPROACH

USGCRP and its participating agencies have maintained a strong portfolio of activities related to observing, understanding, and projecting changes in the physical climate system (see Figure 3.1), and have taken leadership in international programs of research and observation (see Chapter 1). This emphasis reflects the most important research questions, research capacities, and funding priorities of participating agencies in the early years of the Program. Over time, the critical need for research on social and ecological systems has been recognized (NASEM, 2017a; NRC, 1992, 2003, 2012). The Program has made efforts to expand and better support these research areas (e.g., USGCRP, 2012); however, research on the physical climate system remains the dominant focus of USGCRP (see e.g., USGCRP, 2020). The committee believes that meeting the decision needs going forward will require rapidly evolving research related to social and ecological systems, with associated increases in agency support.

Framing research priorities to manage systems-based risks also demands a much more integrated approach to research, indicated in Figure 3.1 for 2030 and beyond. Human actions are changing the dynamics of the natural system in ways that can, in turn, alter human systems, that then further change the dynamics of the natural system, in an ongoing feedback loop. A perfect understanding of individual parts or subsystems does not automatically lead to an adequate understanding of the behavior of the whole Earth system. Such an integrated systems-approach requires a more complete integration of the natural and social sciences than has been achieved by USGCRP to date.

Over the next decade, the grand challenge for the USGRP will be to integrate and communicate knowledge across the physical, ecological, and human systems to provide a more complete understanding of the Earth System and its complexity. Indeed, the most pressing research on the climate-related risks described in Chapter 2 requires knowledge of the integrated Earth System to manage climate change which is the grand challenge for society in the 21st century.

Integrated Systems-Based Research

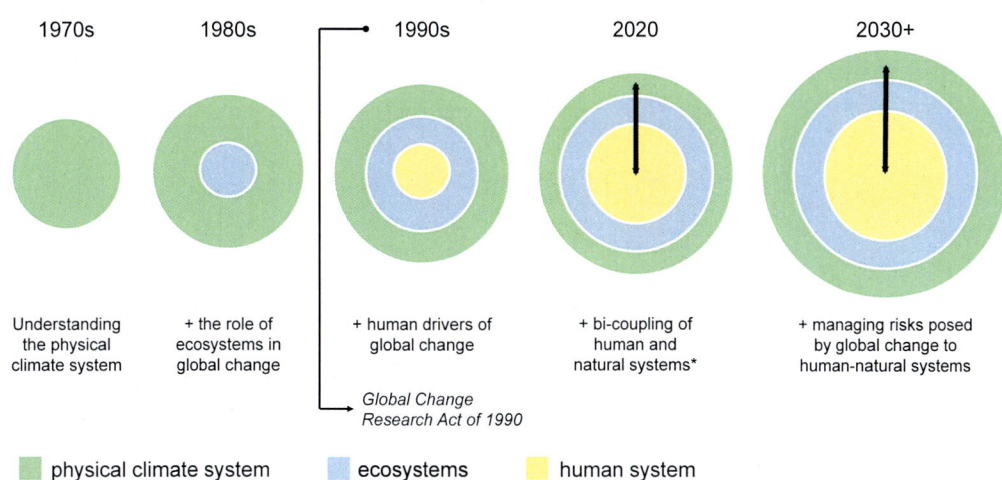

FIGURE 3.1. This figure illustrates the evolution of the scope of global change research since the 1970s, when global change research was largely focused on understanding the physical climate system (green circle). In the 1980s, research expanded to more intentionally include biology and ecology, and to consider the role of ecosystems (blue circle) in global change. In 1990, when the GCRA was made law, human activities were increasingly recognized as major drivers of global change (indicated by yellow circle), as noted in the language of the Act. In the years since the GCRA, understanding of global change has evolved to consider the coupling and bi-directional interactions of human and natural systems (indicated by arrows), where "natural systems" include the physical climate system as well as ecosystems. In 2030 and beyond, it is this committee's contention that global change research (and ergo USGCRP's purview) should also include management approaches to reduce the biggest threats posed by global change to the coupled human-natural system. Note that the increasing size of the circles illustrates the increase in scope of what was broadly considered within the scope of global change research, not to indicate any specific investment or dollar amount that attributed to any of the categories, although additional financial resources clearly will be required.

Given its mandate to coordinate research across multiple agencies and multiple dimensions of global change, it is imperative USGCRP play an important role in accelerating integrated systems-based research and encouraging this approach in cooperative international efforts. The Program has made steps forward since the last decadal plan (notably the sector-based assessments on food security and impacts of climate change on human health). Meeting the urgent decision needs of the next decade will also require a much greater commitment to research efforts that take a whole systems view, involve experts from the social and natural sciences, more explicitly consider the interactions among natural and human systems, and involve stakeholders throughout.

An integrated systems-based risk-management approach will enable USGCRP to more fully meet the mandate of the GCRA, given the urgency of addressing climate impacts happening today and projected risks for the near future. This approach is the logical extension of the research priorities described in the Act, reflecting the progression of knowledge and the advancement of data and tools. Continued advances in fundamental and applied Earth system science over the next decade will be significantly more useful if the integration of natural and social sciences is prioritized.

HUMAN SYSTEM AND HUMAN-NATURAL SYSTEM SCIENCE TO SUPPORT DECISION MAKING

An important goal of global change research is to increase resilience through improved understanding of ways to effectively manage interdependent risks within and across systems and sectors while meeting other societal objectives. There are many dimensions to this challenge. The decision space for risk management includes natural science questions about the magnitude, pattern, and timing of hazards associated with future global change and the responses of ecosystems to global changes. The decision space also includes fundamental social science questions about which regions and populations will likely be more vulnerable to individual and compound risks, and the factors that influence decision making; how knowledge spreads and is taken up across social networks; the degree to which management strategies from one place are useful in another; the appropriateness and availability of technologies; the distribution of risks, costs, and benefits across social groups; and how these could shift over time under different assumptions of climate change and socioeconomic development (Janetos, 2020). There also are economic questions about the costs and benefits of mitigation and adaptation policies and programs, including their trade-offs and synergies over spatial and temporal scales. Because much of the societal context of future worlds is unknown, the shape of future risks is likely to shift with climate change and with policy decisions that may or may not be made in a timely manner (Ebi et al., 2016). More transdisciplinary research into these questions can lead to better informed policy designs and public and private incentives for policy adoption and effective implementation that considers behavioral factors that determine (or undermine) their success.

For example, greater understanding is needed about how future development choices could affect not just greenhouse gas (GHG) emissions but also trends in population growth, urbanization patterns, human migration, economic growth, investments in scientific research, and technology development and deployment. These trends will influence the magnitude and pattern of risks, and the extent to which communities and states will likely invest in mitigation and adaptation to avoid preventable impacts and reduce residual risks. Future development choices have greater impacts on the magnitude and pattern of future risks than climate change alone, particularly until mid-century (Byers et al., 2018). Investments in social science research are needed to improve understanding of the socioeconomic consequences of climate change, as well as behavioral, institutional, and political drivers of climate at different scales—the implications for migration, global security, supply chains, governance, human health, the insurance industry, and a host of other issues that together define the societal consequences of climate change. And, of course, all of these will unfold differently for different social groups.

Understanding of vulnerability and resilience can be gained through studying how changing conditions of populations, places, infrastructures, and environmental, socioeconomic, and political systems interact to affect exposure, susceptibility to harm, capacity to manage, and eventual impacts. These interactions are constantly evolving and influenced by public and private policy (McLaughlin, 2011). This knowledge can be used to project changes in vulnerability that can inform modeling of the magnitude and pattern of risks to provide more robust estimates of possible future challenges under a range of climate and development pathways. The projections can be used to prioritize investments into research and technology to be prepared to address those challenges.

Limiting climate risk also requires greater understanding of the technical potential for reducing emissions and requires equity and ethical considerations (i.e., what is feasible given existing or developing technologies, and at what potential cost to whom). Another key goal is understanding any path dependencies created that would reduce future flexibility in reducing emissions. Technical potential does not directly translate into desired outcomes if technologies are not adopted or relevant behaviors do not change. An effective risk-reduction strategy should facilitate businesses, communities, cities, and households to adopt the technology, take on any needed expenditures, and change behaviors. Boundary spanners, such as agricultural and coastal (e.g., Sea Grant) extension agents, may be needed to facilitate adoption of the technologies through education and training. Adoption, either of emissions reduction activities or of activities that reduce damage, depends on the incentives and constraints faced by households, firms, etc.; these are shaped by public decisions and by public and private initiatives, as well as how they are implemented. Thus, research is needed not only on policy design (and associated incentives) but also on the incentives for policy adoption and effective implementation, the behavioral factors that determine their success, and any path dependencies created (Bidwell et al., 2013; Lemos et al., 2014; Nielsen et al., 2020).

The goal is to identify paths forward that facilitate effective initiatives that individuals and communities are willing to adopt (i.e., that will likely lead to technically feasible and achievable behavioral changes that are desirable from a risk-management perspective [Caniglia et al., 2021; Stern and Dietz, 2020] and that consider synergies and trade-offs). This effort should include research related not only to public policy, but also to private sector initiatives because private firms are increasingly influencing climate-related choices by individuals and throughout supply chains (Gilligan and Vandenbergh, 2020; Vandenbergh and Gilligan, 2017). Additionally, nonstate civil society organizations and nonprofits are also a means to collective action (Tosun and Schoenefeld, 2017). USGCRP and the National Climate Assessment could catalyze directly the development of sustained public-private collaborations that can connect appropriate expertise to decision makers in federal, state, local, and tribal governments to ensure adoption of acceptable (to affected groups), humane, resilient, and equitable adaptive measures (Moss et al., 2019; NASEM, 2016a).

Additional research needs to support understanding the decision space for risk management include the following:

- The possible path dependencies, synergies, and trade-offs between mitigation and adaptation over time, both the physical interdependence where an action affects emissions reduction and aids resilience, and socioeconomic factors that influence public support and action, such as addressing environmental justice in energy transitions.

- Projections of how changes in hazards, exposures, and vulnerability over temporal and spatial scales could shape future risks and resilience.

- Applications of exploratory modeling that deal with deep uncertainty through iterative modeling processes that analyze the implications of different potential solutions across diverse potential futures (e.g., Moallemi et al., 2020; see Box 5.1).

- Potential responses of households, firms, public agencies, etc. to public and private initiatives intended to spur emissions mitigation, and the design of programs and policies for maximum effect.

- Use of an equity focus to promote resilience and sustainability. Climate change is exacerbating current and creating new inequities. Historical inequities drive current inequities that drive future inequities that in turn drive exposure and vulnerability. Understanding these linkages could improve the ability to address future inequities, exposures, and vulnerability.

- Improvement of science communication to more effectively provide global change data, information, and tools for a range of stakeholders and risk-management decisions (NASEM, 2017a).

DESIGNING AND IMPLEMENTING INTEGRATED SYSTEMS-BASED RESEARCH

Traditional approaches to designing and implementing global change research are likely to yield stove-piped, science-driven research programs similar to those in place. New approaches to setting research priorities are needed that put user needs at the forefront; doing so would attract a broader and more diverse set of stakeholders and incentivize integrated research. The next strategic planning process will need to embrace and incorporate these approaches more fully, taking advantage of public participation scholarship. Likewise, the planning process will need to engage a much broader swath of federal agencies and other partners to more fully meet needs for global change information and to take best advantage of diverse capacities.

User and Public Participation in Global Change Research

Making global change science useful for effective and timely decision making that (1) protects communities and assets in the short term and (2) fosters needed resilience over the coming decades will require ongoing discourse between the research community and those who are concerned about, are impacted by, and make decisions that influence global change (NRC, 2010a). This needs to be done while acknowledging the significant uncertainties over what changes will prove to be most critical when communities and systems experience impacts. Because researchers do not always understand or prioritize the needs of stakeholders and decision makers, users should be involved in setting the research agenda and, in some cases, actually be involved in designing and conducting research. Advantages include that participatory approaches build trust in the science, help the research community calibrate analysis to local contexts by drawing on indigenous and local knowledge, and help direct scientific attention to issues and questions that will influence risk-management decisions (McClymont Peace and Myers, 2012; Ziegler et al., 2019). Users who are involved in setting, implementing, and communicating a research agenda are more likely to embrace its findings and information products (see, e.g., Gunderson and Dietz, 2018; O'Grady, 2020).

The numbers of decisions that need to be made at local to regional scales place new demands for knowledge generation at relevant, actionable scales. USGCRP and its participating agencies should purposefully include as users highly diverse segments of the U.S. and global populations, including racial and ethnic minorities, socioeconomically disadvantaged people, environmental justice communities, and others (e.g., Dietz et al., 2020; see also Chapter 5 section *Diversity, Equity, and Inclusion in Global Change Research*). Integrating user needs into the research agenda means more closely coupling user needs into how the USGCRP research agenda is set. Engagement with the larger body of interested and impacted parties, not just government agencies, is essential to assure that the research is broadly useful and attentive to local, regional, and sectoral contexts. Understanding these contexts is crucial, as is the need to ensure that equity-based solutions are central to the implementation of research findings.

USGCRP has laid the groundwork for linking analysis and public deliberation in the National Climate Assessment (e.g., through stakeholder engagement in development of technical inputs and participation on author teams). Over the next decade, USGCRP will need to place strong emphasis on this approach. Of special note is the necessity to engage with underserved and disadvantaged communities, who are often at especially high risk from global change, and to invest in research on the deliberative processes themselves so as to learn from experience and enhance capabilities.

Disciplinary integration in global-change research is also expected to improve the utility of research outputs to user communities. Transdisciplinary efforts like participatory scenario development require model and other research outputs that are usable by a diverse audience, and they invite an exchange of knowledge rather than a one-way transfer. USGCRP could expand opportunities for mainstreaming findings into user communities through these sorts of transdisciplinary approaches, which increase the likelihood that information communicated will be appropriate, timely, and useful.

Previous National Academies of Sciences, Engineering, and Medicine reports on climate change and other complex issues at the interface between science and decision making have called for approaches that link scientific analysis to an ongoing public deliberation (NRC, 2010a, b, 2011). National Academies reports have also pioneered the theory and evidence behind such approaches, providing guidance on how they can be implemented (NRC, 2008). The approach is now being called for to handle many science-based policy issues (Dryzek et al., 2020; NASEM, 2016b, 2017b). The committee emphasizes that the design of the exact mechanisms appropriate for such engagement needs to be tailored to the circumstances, but recognize that ongoing research on participation has led to design principles that can provide guidance (NRC, 2008).

CHAPTER FOUR

Research on Approaches Critical to Managing Climate Risk

> **RECOMMENDATION: Prioritize research related to managing climate risks, including: (1) reducing global greenhouse gas emissions and lowering their atmospheric concentrations; (2) increasing resilience to current and anticipated climate-related security risks; and (3) expanding research on incentives for and the synergies and trade-offs between these risk-management approaches.**

The urgency of the rate and magnitude of climate change and the complexity of interactions across risks and responses mean that the next decade will require immediate investments in coordinated research to protect human systems and ecosystems from the risks described in Chapter 2. Research is required to support decision making that integrates climate risk-management strategies and policies. The primary strategies are (1) mitigation, reducing global emissions and removing CO_2 from the atmosphere; and (2) adaptation, preparing for and managing the harmful effects of global change. These strategies can be reinforcing or have unintended, and potentially negative, consequences. Solar geoengineering—the deliberate large-scale manipulation of an environmental process that modifies the amount of solar heating of the Earth—is another strategy that requires further research to better understand its technical and social feasibility, as well as how such measures could interact with mitigation and adaptation in ways that may introduce additional risk.

Because adaptation was extensively covered in Chapter 2, the primary focus of this chapter is on mitigation and solar geoengineering. In addition, Chapter 3 introduced important social science needs required to understand how to achieve emissions reductions for policy design and through behavior change that include considerations of ethics, equity, technical potential, adoption, and path dependencies.

Many strategies for reducing emissions or removing carbon from the atmosphere have implications for the kinds of risks noted in Chapter 2, underscoring the need for an integrated research approach. In some cases, a technological solution may affect natural systems, such as the potential for solar geoengineering to change the levels of ultraviolet light received by plants, with potential implications for agricultural productivity and ecosystem health (NRC, 2015). In other cases, there may be opportunities to make natural systems more resilient while also limiting the overall impact of climate change, for example strategies to enhance storage of carbon in natural systems are often similar or identical to strategies that improve soil health and water retention in the landscape (NASEM, 2019a).

Although one objective of the 2012 Strategic Plan recognized the need to "enhance the usability of scientific knowledge in supporting responses to global change" (USGCRP, 2012, Objective 1.2: Science for Adaptation and Mitigation), to date the program has not included a major emphasis on understanding the effectiveness of these risk-management strategies or associated benefits, costs, and equity impacts. Yet, such an understanding is an essential input into decision making at multiple levels.

REDUCING RISK BY GLOBAL EMISSIONS REDUCTION

The ultimate requirement to avoid additional climate change risk involves reducing net anthropogenic emissions of the forcing gases to zero. In 2030, the United States is projected to emit only 11 percent to 12 percent of global emissions of fossil and industrial CO_2, the dominant greenhouse gas (GHG), with this share projected to decrease over time (IEA, 2020). Thus, in addition to understanding the effectiveness of emissions reduction strategies within the United States, it is critical for the U.S. Global Change Research Program's (USGCRP's or "Program's") research efforts to consider how U.S. actions and decisions can most beneficially affect those of other countries. This influence comes from active U.S. participation in the cooperative programs referenced in Chapter 1, which is usefully supplemented by additional effort to develop techniques to measure GHGs on national and smaller scales to build confidence in national emissions pledges. Difficult-to-measure emissions from soils deserve special attention. Other crucial channels of influence include leadership in the global institutions that coordinate discussions on the control of GHG emissions, formulate international climate developments, and marshal financial and technical aid for developing regions. Climate action within the private sector is a vital part of this engagement, because supply chains span the globe, and standards and practices adopted by firms in one nation often influence those employed elsewhere.

Here, the committee focuses on USGCRP's mitigation research in three areas: setting science-based mitigation goals, improving emissions measurement and monitoring, and exploring a range of CO_2 removal and sequestration techniques. Many mitigation measures provide a joint benefit in increasing the security of particular U.S. sectors through a simultaneous contribution to efforts to adapt to a changing climate. However, also important to consider is the range of risks associated with different mitigation approaches. For example, CO_2 removal and sequestration technologies themselves pose various and substantial risks and uncertainties, with respect to water, energy demand, land-use and costs that require research to better understand their efficacy and potential to reduce net emissions in a changing climate (NASEM, 2019a).

Enhance the Scientific Basis for Mitigation Goals

The choice among widely discussed global climate mitigation goals (e.g., limiting global warming to 1.5 or 2°C) needs to be strongly informed by science (IPCC, 2018). For example, recent emphasis on limiting global warming to 1.5 rather than 2°C arose in part from improved understanding of how much sea level is likely to rise at 2°C of warming over what timescales, particularly from the contributions of large land ice sheets in Greenland and the Antarctic. That understanding, however, is far from complete, and as it is refined, policy goals may be altered.

Informing top-line mitigation goals (e.g., limiting warming to 1.5 or 2°C) requires understanding the sensitivity of climate to human forcing and the physical manifestations of different levels of emissions; the environmental and socioeconomic consequences of climate change under different levels of mitigation; and the socioeconomic consequences of the mitigation (and adaptation) strategies for different populations and regions, including implications for equity. In turn, the effects of social change on GHG emission and land use, including the effectiveness of programs and policies intended to reduce emission, need to be considered in scenarios used for modeling. Acknowledging there is sufficient information to take urgent action now (Gilbert and Sovacool, 2016; Wara, 2015; Wara et al., 2015), this fuller understanding of the global societal consequences of different levels of warming could inform additional actions to refine and meet mitigation targets. Achieving fuller understanding of the socioeconomic consequences of climate change is a major motivation for greater integration of multidisciplinary research, particularly natural and social sciences. Some consequences are amenable to sector-by sector treatment, whereas others, such as migration pressure and possible political instability, are more crosscutting.

Understanding the physical and socioeconomic implications of different levels of global warming is also foundational to adaptation planning. For that purpose, this information needs to be produced at the local spatial scale.

Improve Emissions Measurement and Monitoring

Accurate measurement and reporting of anthropogenic emissions of GHGs at the national scale is foundational to controlling global GHG emissions. This means measuring emissions from fossil fuel burning and from direct human intervention in the land sector (land use and land-use changes), such as deforestation.

In the context of the United Nations Framework Convention on Climate Change (UNFCCC) reporting, nations are encouraged to use GHG emissions-measurement methods issued by the IPCC (2006, 2019c). For many sectors, a hierarchy of methods (of increasing complexity and accuracy) is presented. These methods emphasize "bottom-up" approaches based on knowledge of human activities. For example, fossil fuel emissions are estimated based on reported fuel usage together with information about the carbon content of the fuel. These are distinct from "top-down" approaches that are based on measured changes in concentrations of CO_2 in the atmosphere. Both approaches are needed.

Emissions are estimated by the emitting nations themselves and reported to the UNFCCC. The current international agreement provides for independent review by technical experts of self-reported emissions. Even so, confidence in self-reported emissions is limited by a general lack of methods and data to independently verify them. The accuracy of reported emissions is limited by factors including lack of capacity for making estimates (particularly in the developing world) and data gaps, and by incentives to report inaccurate emissions for political, financial, or economic purposes.

Research on scientific measurement of GHG emissions should identify methods that will be effective in the context of likely policy approaches to emissions verification. Many of the recommendations in previous National Academies reports (NASEM, 2018; NRC, 2010c) on characterizing and verifying emissions are still relevant today, including: maintaining essential surface and satellite observations networks, supporting data assimilation systems, expanding gridded inventories, and coupling top-down and bottom-up approaches.

Explore CO_2 Removal, Reliable Sequestration, and Utilization

Mitigating GHG concentrations necessarily involves cutting global GHG emissions. This response may be supplemented, however, by activities that remove CO_2 from the atmosphere and either store it or convert it to some other form (NASEM, 2019a). Research on the mitigation of domestic U.S. emissions—including the development of technologies to aid replacement of fossil fuels, and federal, state, local, and tribal control measures—is ongoing in agencies outside the normal scope of USGCRP activities. CO_2 removal, on the other hand, falls outside purely domestic efforts, and merits treatment within a wider USGCRP concern with issues of global scope. Applications of direct removal technology may involve similar multination efforts. As mentioned above, different carbon removal and storage approaches come with different associated risks to the systems identified in Chapter 2 (e.g., food, water, and energy), which also require attention and research in the context of other approaches.

Additionally, only a small fraction of the CO_2 and methane emitted each year is currently being captured and used, and most technologies to utilize captured carbon are in their infancy. However, these technologies have a role to play in future carbon management and the overall portfolio of mitigation strategies. A recent National Academies report (NASEM, 2019b) found that carbon utilization needs to be done at scale, which will depend on the pace of technology development and future energy, market, and regulatory landscapes. The report also found that, like all technologies, "a comprehensive evaluation of carbon utilization technologies would include evaluation at various maturity levels based on economic, market, regulatory, and environmental factors" (NASEM, 2019b, p. 4). There are other integrating factors across human and natural systems of global change to be considered in further technology development that may involve social or regulatory barriers and incentives as well as disruptive change to energy and material manufacturers. Better integration of current research efforts is needed to advance progress in this space (NASEM, 2019b).

ADAPTATION TO REDUCE RISKS

Climate change currently affects the security of the American people and the nation across many systems including human health, food, water and energy, with projections concluding that, without considering adaptation, each additional unit of warming would further increase nearly all risks with the risks differentially affecting different ecosystems, regions, and human populations. Adapting to these risks has been on USGCRP's research agenda for the past two decades. However, new research, enhanced coordination, and expanded communication efforts are needed to advance society's ability to adapt to risks arising sooner and more intensely than projected, within the context of increasingly complex interactions among these security risks (Janetos, 2020). Further, longer-term evaluation is needed to monitor the effectiveness of adaptation options over time to identify adjustments needed to enhance resilience.

Advancing an integrated understanding of security risks was the focus of Chapter 2. This integrated understanding is an important component of research needed to inform efforts to adapt to climate change. In addition, the research of USGCRP and participating agencies will increasingly require engaging in ongoing discussions with decision makers to support specific adaptation decision needs. Examples of these sorts of research questions are provided in Box 4.1 for coastal communities.

Box 4.1
Integrating Case: Managed Retreat for Coastal Communities

The idea that coastal communities could retreat inland is one possible adaptation to climate change risks. For some communities, retreat may be necessary or the communities may have already self-identified the need to relocate (Dannenberg et al., 2019). Retreat/relocation options need to be examined broadly, including economic and socio-cultural as well as physical factors, and with intent to move from a piecemeal to a strategic approach. Careful examination of the many ramifications of potential community relocation away from the coast is essential to inform assessments of whether it is a good option for a particular situation (Siders, 2019). Assessments will require:

- Better understanding of climate and coastal dynamics, socioeconomic changes, and ongoing adaptation efforts (e.g., Siders et al., 2019; Siders and Keenan, 2020);
- Results of prior property buyouts and other relocations (e.g., Mach et al., 2019);
- Consideration of monetary and socio-cultural costs and benefits, including those that inland communities may incur to accommodate relocating people (e.g., Clément et al., 2015); and
- Much greater attention to community involvement, equity, vulnerability, environmental justice, transparency, and policy requirements (e.g., Schlosberg and Collins, 2014; Siders et al., 2019; Siders and Keenan, 2020).

In addition, the Federal Emergency Management Agency, which is responsible for elevation mapping, designation of flood-prone areas, the National Flood Insurance Program, and the purchase of flood-prone housing, should enhance incorporation of sea level rise projections in its flood guidance to states and communities and work with states to improve retreat/relocation options. However, some municipalities are already taking steps to incorporate sea level rise in local building regulations (e.g., the town of Mount Pleasant, South Carolina, is raising the elevation required for new construction).

Siders et al. (2019, p. 763) concluded: "A substantial amount of innovation and work—in both research and practice—will need to be done to make strategic, managed retreat an efficient and equitable adaptation option at scale." USGCRP research could make a substantial and needed contribution to scaling up knowledge of coastal adaptation if it is framed broadly, including ethical, economic, sociocultural and physical components of the issue.

SOLAR GEOENGINEERING APPROACHES

"Solar Geoengineering" approaches aim to limit climate change through a variety of climate interventions that modify the amount of solar heating of the Earth. Not enough is known about these approaches and their potential impacts to consider deploying them today, but concerns that mitigation and adaptation efforts will be insufficient to avoid the worst consequences of climate change have motivated a call for increased research on solar geoengineering in case it is needed in the future (NRC, 2015).

Possible solar geoengineering approaches include widespread distribution of small reflective particles in the stratosphere, augmentation of reflective cloud cover in the lower atmosphere, or reduction of cirrus clouds in the upper troposphere that trap outgoing radiation. While these approaches could potentially reduce global atmospheric temperature and reduce some near-term risks of climate change, they could also introduce new risks—such as reduction in stratospheric ozone, shifts in precipitation patterns, or impacts on ecosystems—with potential implications for geopolitical instability (NASEM, 2021b). Indeed, even research on the feasibility of solar geoengineering has raised concerns about the "moral hazard" involved—specifically the notion that holding out the promise of these options in the future might forestall efforts to mitigate and adapt to climate change now (NASEM, 2021b).

A 2015 National Academies report anticipated several scenarios in which it would be beneficial to understand the risks and opportunities involved with sunlight reflection strategies. That report recommended that research on these strategies be expanded and that a serious deliberative process be undertaken to examine what sort of research governance is needed—emphasizing that open conversations, with civil society engagement, should be part of the process of oversight for any research efforts undertaken. However, investments in this research are still very modest.[1]

The National Academies launched a consensus study in 2019[2] to develop a detailed transdisciplinary research agenda and recommend research governance approaches for solar geoengineering. The study committee was tasked to identify a wide range of research needs spanning feasibility, efficacy, and risks; to provide detailed guidance on research design and research governance; and to discuss mechanisms to ensure

---

[1] See https://geoengineering.environment.harvard.edu/blog/funding-solar-geoengineering.
[2] See https://www.nationalacademies.org/our-work/developing-a-research-agenda-and-research-governance-approaches-for-climate-intervention-strategies-that-reflect-sunlight-to-cool-earth.

transparency, accountability, and legitimacy of the research outcomes. The recommendations from the National Academies consensus report on this topic (NASEM, 2021b) are relevant to USGCRP's mission and mandate and should be carefully considered by the Program.

Research on solar geoengineering spans the range of physical, ecological, and social sciences that contribute to global change research. To be successful, such research will also need to coordinate a number of stakeholders in the U.S. scientific community, including USGCRP participating agencies, as well as to foster collaboration and consideration of relevant international programs and activities. As such, the committee sees the potential for USGCRP to play a pivotal role in advancing this research, particularly if the Program takes steps to improve both its disciplinary representation and its efforts to engage stakeholders in defining research priorities.

A NEED FOR INTEGRATED RESEARCH ON RISK-MANAGEMENT APPROACHES

Integrated systems-based research is urgently needed to describe and quantify the complexities of interactions across sectors, regions, and decision-making entities, considering the interdependence, synergies, and trade-offs among mitigation, adaptation, solar geoengineering, and strategies to address other societal priorities. For example, many risk-reducing actions decrease both the human activities driving change and the damage from climate change that cannot be stopped.

CHAPTER FIVE

Crosscutting Research and Data Priorities

> **RECOMMENDATION: Expand research in five crosscutting areas: (1) extremes, thresholds, and tipping points; (2) regional- and local-scale climate projections; (3) scenario-based approaches; (4) equity and social justice; and (5) advanced data and analysis frameworks.**

Basic understanding is needed to inform mitigation and adaptation policies. For example, understanding is needed of the full physical and socioeconomic consequences of different levels of global warming. Physical consequences include, for example, changes in weather and climate extremes, and socioeconomic consequences include those discussed in Chapter 2. This understanding is needed to set an appropriate top-line mitigation goal (e.g., limiting warming to 1.5 or 2°C) and also to anticipate and plan how to live in a world with that level of warming. Such understanding is foundational to the National Climate Assessment, which takes on new importance as climate change impacts become material and remains a legislative mandate. All of this requires improved understanding of not only fundamental processes, but also of local-scale physical manifestations, as well as granular assessments of consequences for ecosystems and human systems.

Closely related, and similarly fundamental to understand, are nonlinear responses to increasing greenhouse gas (GHG) emissions, including thresholds, tipping points, and physical and carbon-cycle feedbacks. These issues directly inform socioeconomic impacts of climate change, as well as more profound questions such as the carbon emissions budgets associated with different levels of global warming, and even the possibility of uncontrolled GHG emissions from biotic sources.

Improved capabilities for modeling the physical climate system remain a key focus of the research coordination efforts provided by the U.S. Global Change Research Program (USGCRP or "Program"); accomplishments from this effort were highlighted by this committee (NASEM, 2017a). However, advances in process understanding, machine learning, scientific computing hardware, and more create the opportunity for significant advances. Specific goals should include better simulation of local-scale phenomena as well as climate and weather extremes, and advances in uncertainty quantification. This requires a suite of modeling tools, including high-resolution earth system models (ESMs), as well as models of intermediate complexity that can be better suited to exploring issues such as uncertainty quantification.

This chapter focuses on crosscutting research priorities that would facilitate an integrated systems-based approach to risk management (which would enhance management of security challenges) including greater understanding of extreme events and tipping points; improved simulation of local and regional-scale climate; the use of scenarios to project possible combinations of climate and socioeconomic development within which security challenges will arise and be managed; augmentation of data and analyses facilities; and a focus on equity and justice issues.

These integrated systems-based approaches to understanding would be facilitated by capitalizing on investments in research from international organizations and institutions, such as the European Union and the World Climate Research Program. The output of the suggested research investments would be critical input to National Climate Assessments, coordination with international assessments, and risk communication. This report assumes these activities will continue to be central to USGCRP.

Furthermore, because communication of research to a range of stakeholders and for a variety of risk-management decisions is central to the mission and practice of USGCRP, the report assumes that the communication of risks and responses will be embedded in all aspects of the Program's next decadal research plan, building on the conclusions from multiple National Academies of Sciences, Engineering, and Medicine reports (NASEM, 2017b; NRC, 2010a). Responding to the demand for data, information, and tools that are credible, comprehensive, useful, and usable to enable decision makers at different scales to prepare for and manage climate change provides the basis for an effective national capacity for managing the risks of and responses to climate change. The national capacity needs to be structured to learn from successes and failures, to share lessons learned and best practices, and to reduce unequal burdens on any one region, sector, or population group.

EXTREME EVENTS, THRESHOLDS, AND TIPPING POINTS

Possible Earth-system responses to human GHG emissions include not only gradual trends, but also increases in certain types of extreme events and nonlinear responses, including some that are self-reinforcing. The gradual trends include continuous, incremental increases in atmospheric levels of CO_2 and other GHGs, ocean heat content, sea level rise, and other environmental variables. Increases in some categories of extreme events, such as extreme heat, extreme precipitation, and wildfire, are well documented (see, e.g., Kossin et al., 2017). The trends for other event types, such as tropical cyclones, are more complicated. Since 1975, the proportion of Category 4 and 5 hurricanes has increased at a rate of ~25–30 percent per °C of global warming (Holland and Bruyère, 2014). This has been balanced by a similar decrease in Category 1 and 2 hurricane proportions, leading to the development of a distinctly bimodal intensity distribution. This global signal is reproduced in all ocean basins. Understanding these extremes is an urgent priority because they have disproportionate societal impacts and can help to inform climate mitigation goals.

The *Climate Science Special Report of the Fourth National Climate Assessment* (Kopp et al., 2017) makes climate tipping points—large-scale, nonlinear shifts in Earth systems—a major focus of its final chapter. Lenton et al. (2019) summarize the current state of understanding of climate tipping points, discuss some of the limitations of the concept, and explore the potential for a cascade of tipping points, with each one triggering others and creating a shift to a warmer world.

Tipping points exist not only in the physical climate and ecological systems, but also in social systems. For example, the introduction of technology led to overfishing of cod and other species in the North Atlantic, resulting in fishery collapse with significant impacts on livelihoods and communities (Hamilton et al., 2004). Of particular concern for the USGCRP research agenda are the interactions between physical and social systems that can lead to surprises and unexpected tipping points with large impacts, such as the drying trends that adversely affect agricultural yields interacting with internal migration, limited employment possibilities, and poor governance resulting in conflict and external migration.

The 2012 Strategic Plan called for USGCRP-supported research to improve modeling of extreme events and identified the possibility of tipping points in physical and biological systems as potential research topics of importance. By the time of the release of the Fourth National Climate Assessment (USGCRP, 2017), advances were made in modeling extreme events, and the assessment included a full chapter on tipping points. However, experience over the past decade and even the past few months highlight the need for continued and expanded research in coupled human-natural system disruptions and socioeconomic consequences of extreme events, tipping points, and social tipping points that might influence mitigation of and/or vulnerability to climate change. Tipping points can also be potentially leveraged for constructive shifts in social-environmental interactions toward low-carbon futures (Otto et al., 2020).

Extreme Events

People will experience climate change primarily through extreme events, many of which are increasing in frequency and intensity, that lead to more compound events with less recovery time in between. Whether or not an event becomes a disaster is a function of the interaction of exposure to the hazard and the degree to which individuals, populations, ecosystems, or infrastructure are sensitive and capable of responding with coping mechanisms, avoidance, adaptation, or transformation (IPCC, 2012; Kim et al., 2020). The vulnerability of interconnected and highly dependent infrastructure, such as water delivery systems that rely on electrical power or emergency shelters that depend on transportation and communication systems, can be amplified by these interdependencies.

The coincidence of multiple types of extreme, climate-related events can compound challenges for communities and regions. Examples include a heatwave experienced coincident with an extreme drought or wildfire, king tides (extreme high tides) coincident with storm surges from coastal storms, and extreme inland storms causing extensive erosion of soils made bare by massive wildfires, in turn driven by the convergence of drought and pest outbreaks. These challenges—most underlain by climate change but many compounded by poor management or inappropriate design—require new preparation paradigms and resilience-building solutions that recognize the uncertainty of where, when, and with what intensity future extreme events will occur as well as that the magnitude and pattern of impacts will be shaped by the vulnerabilities and capacities of exposed communities. One thing appears certain: the history of previous extreme events is now a poor guide to likely future occurrences.

Extreme events also tend to exacerbate existing fissures in society and deep structural inequities. Responses to extremes events can also be considered opportunities to address such structural inequities and rebuild impacted systems to not only be more resilient but more sustainable in the future. New research should further develop understanding about how individuals, organizations, communities and governments assess the likelihood of, perceive, and respond to extreme events.

Tipping Points

One of the lessons of the COVID-19 pandemic is that a potentially manageable crisis can create societal tipping points as impacts on one system (e.g., health) cascade to affect economic and other systems. The COVID-19 pandemic provides important lessons for what can happen when a sudden and unexpected change occurs for which there was insufficient preparation and planning. Slow and inconsistent implementation of required interventions, including adequate testing, contact tracing, physical distancing requirements, and other measures, resulted in community spread, millions of cases, and hundreds of thousands of fatalities in the United States alone.

The cumulative impact of incremental changes in weather and other environmental variables could, at some point in time, push a part of the Earth system beyond a tipping point and into a completely new state. Two recent Intergovernmental Panel on Climate Change Special Reports (IPCC, 2018, 2019b) suggested that some tipping points could be exceeded with just another 0.5 to 1°C of warming, with an increase of 0.5°C projected to occur as early as the end of the next USGCRP decadal strategy.

It is important to note that the potential consequences of geophysical tipping points will depend on the resilience of social systems, with the potential for impacts to cascade through economic, agriculture, water, energy, and health systems. Tipping points in social systems can arise before geophysical tipping points in situations with high vulnerability. Social systems are themselves subject to tipping points that can lead to widespread and rapid social changes (Otto et al., 2020; Shwom, 2020; Smith et al., 2020). Given the widely acknowledged need for rapid change in technology and practices to meet the challenge of climate change, USGCRP should give some priority to research that deploys the substantial literature on rapid social change to provide insights into strategies that might quickly enhance uptake of mitigation and adaptation research.

Tipping points can form a cascade, with each one triggering others to create an abrupt shift in the planet's climate system or in social systems that is irreversible. An example of a tipping point cascade involves the ocean circulation system that moves heat around the planet and plays a key role in climate patterns (Steffen et al., 2020). Greenland ice in a warmer Arctic drives a key component of ocean circulation to a 1,000-year low. Fresh water from the melting flows into the Labrador Sea, which has the potential to increase the buoyancy of surface waters and reduce formation of dense, deep water that helps drive the overturning circulation (Yang et al., 2016). Further decline in the ocean current in the Atlantic could lead to a shift in heat distribution around the planet that could trigger other tipping points. Potential tipping-point cascades should be investigated using a variety of tools; USGCRP needs to invest both in high-complexity, high-resolution earth system models (ESMs) and lower-resolution, faster ESMs of intermediate complexity that allow more thorough exploration of uncertainty. To the extent that there is a trade-off between achieving higher resolution and allowing more simulations to better understand uncertainty, those trade-offs can be assessed based on the value of the information generated and its

GLOBAL CHANGE RESEARCH NEEDS AND OPPORTUNITIES FOR 2022-2031

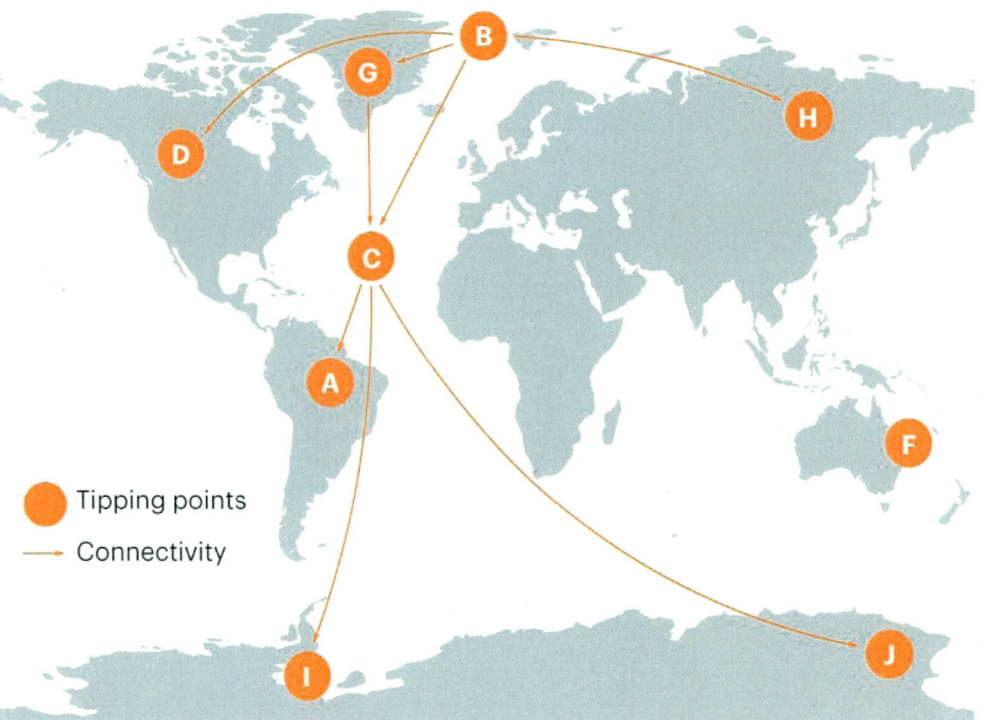

FIGURE 5.1. More than a dozen potential tipping points are being monitored, including losses of ice mass from the ice sheets in Greenland and the Antarctic; losses of terrestrial carbon to the atmosphere as CO_2 and/or CH_4 from ecosystems that span the full latitudinal gradient from tropical forests to tundra ecosystem; shifts in major seasonal weather patterns such as the monsoons in Asia; thermal stress of coral reefs; and disruption of the ocean circulation system. Source: Lenton et al., 2019.

likely impact on decisions. Equally important is the need for expanded programs of observations to form the basis of improved representation in models of physical processes leading to potential tipping points. Such processes include permafrost thaw and associated GHG emissions and response of the terrestrial and ocean carbon sinks to climate warming.

Finally, it is important to continue to recognize that in addition to extreme events and tipping points, there is significant potential for humanity's effect on the planet to result in unanticipated surprises. Kopp et al. (2017) noted that there is a broad scientific consensus that the further and faster the Earth system is pushed toward warming, the greater the risk of such surprises.

Selected Needs to Meet These Challenges in 2030

Assessment of the likelihood and timing of extreme events and abrupt changes, including cascading changes, has been difficult to quantify due to insufficient data and a limited ability to model the underlying physical and biological processes. As a consequence, it is difficult to account properly for extreme events and the possibility of major changes in the Earth system in risk projections (van Ginkel et al., 2020). The socioeconomic impacts of the COVID-19 pandemic may provide a useful model for considering the types and magnitudes of such adverse effects, including those on marginalized and at-risk populations, and the need for targeted research investments (Bonaccorsi et al., 2020; Nicola et al., 2020; Singu et al., 2020).

To improve projections of the likelihood and timing of extreme events and abrupt changes in the climate system and the magnitude of consequences for society, three related research efforts could be undertaken. First, existing ESMs should be modified to facilitate simulations of individual tipping point cascades. Second, capacity should be developed to model the two-way associations between climate and social tipping points, including economic shocks and societal disruptions such as forced migration and environmental justice disparities. Third, integrated assessment modeling efforts should be advanced to develop the capacity to link climate and social tipping points (van Ginkel et al., 2020).

SIMULATION OF LOCAL- AND REGIONAL-SCALE CLIMATE

Decisions to manage the risks of climate change need to be informed by projections at relevant scales. The spatial resolution of global climate models (GCMs) continues to improve but is still generally insufficient to directly inform most decisions. This needs to be addressed through development of much finer-resolution GCMs and improved "downscaling" of GCM results. Furthermore, GCMs do not simulate or do not simulate well many climate-related hazards, including wildfires, floods, tropical cyclones, and tornadoes. Improved representations of these phenomena are needed, whether within climate models or—as is common practice—in separate models driven by climate model output.

In general, downscaled projections result from either dynamical or empirical downscaling approaches. Dynamical downscaling refers to techniques that rely on dynamical climate models, either regional climate models or variable resolution models. Empirical statistical downscaling relies on developing relationships between large-scale variables produced by reanalyses of past climate data or GCMs (e.g., 500 mb heights) and local variables (e.g., temperature and precipitation) needed for impacts and adaptation work, very often at the single point scale. Then these relationships are used to determine changes in the local variables for the future period. Each approach has strengths and weaknesses; these often determine which approach is used, but often the decision is pragmatic rather than scientific. For example, statistical downscaling approaches are less computationally expensive than dynamical methods; thus, it is easier to downscale a large number of GCMs. Dynamical methods, on the other hand, more easily provide a large suite of variables (e.g. winds, humidity, snow) and thus are useful for studies that require more exotic variables. An important development that needs to be further pursued in dynamical downscaling is the value added of developing climate projections at convective resolving scales (Prein et al., 2015).

Numerous comparisons (e.g., Tang et al., 2017) indicate that the methods project different climate changes, based on the same GCM, particularly for precipitation. However, why they differ has not been sufficiently explored and thus which method would be more credible in a particular region is not easy to determine. More rigorous comparisons are needed of dynamical and statistical approaches.

According to a 2012 National Academies report, "Although different approaches to achieving high resolution in climate models have been explored for more than two decades, there remains a need for more systematic evaluation and comparison of the various downscaling methods, including how different grid refinement approaches interact with model resolution and physics parameterizations to influence the simulation of critical regional climate phenomena" (NRC, 2012, p. 71). USGCRP should build on its strong track record in advancing ESMs and coordination of different U.S. modeling centers through its interagency working group.

Socioeconomic Downscaling

The downscaling of socioeconomic aspects of scenarios also is critical. In this context, downscaling encompasses providing both scenarios and the empirical data needed to calibrate and assess scenarios at finer scales, including communities and groups within communities that may be differentially impacted by climate change. With regard to the data needed for verification, ultimately data on individuals and households is ideal although more aggregate data on small areas or local jurisdictions can often be useful when finer grain data is lacking.

Downscaling of variables beyond aggregated measures of demographic and economic change, such as measures of equity and of patterns of urbanization, are needed by decision makers. Quantification and downscaling are needed of variables relevant for adaptation, such as measures of extreme poverty, quality of governance, water scarcity, innovation capacity, extent of social protection, and educational attainment (Schweizer and O'Neill, 2014).

Local- and regional-scale simulations are also needed to quantify some of the co-benefits to society in specific locations associated with mitigation. Examples of quantification of potential co-benefits include climate effects on local air quality, including particulate matter loads, and the potential for carbon sequestration associated with afforestation.

Scientific findings can be made more readily actionable at decision-relevant scales when projections are informed by the range of urgent challenges faced by local or regional entities. For example, urban practitioners are calling for downscaled climate information to understand the likelihood of future extreme events, and their interactions with the built environment, along with projections of possible future distributions of vulnerable populations to facilitate planning for implementing defensive strategies. Coastal states need information about sea level rise, coastal storm frequency and intensity, and socioeconomic projections such as population growth, property values, and availability of insurance to plan for coastal realignment or fortification. To wisely manage future water resources, municipalities dependent on water supplies from distant watersheds need information on future climate, hydrology, and ecology of those watersheds, plus projections of population and business growth in their municipality. These examples suggest the importance of ongoing dialogue between the research community and those who need the emerging understanding to inform their decisions. In addition, research is needed to clarify what degree of spatial and temporal resolution is useful for decision making as many decisions may not require high resolution.

SCENARIOS-BASED APPROACHES

The term *scenario* is used in multiple contexts, with different meanings. Scenarios do not predict the future but facilitate exploration of a range of possible futures, their associated risks, and the extent to which mitigation and adaptation could reduce projected risks. Scenarios are used to explore what could happen under different sets of assumptions. Approaches to scenarios for global change research include (1) models based on internally consistent descriptions of how future drivers of GHG emissions could evolve over the course of this century; and (2) participatory-based descriptions of factors that can inform specific policy development.

To be effective in identifying research needs, scenarios should be targeted to the full range of scales—global, national, regional, municipal, and community-based—so that USGCRP research can be integrated into the mitigation and adaptation policy portfolios of decision makers and legislative bodies, enabling them to explore potential conflicts, trade-offs, and synergies. Use-inspired research[1] to inform such decisions is also needed on coupling downscaled climate models with multiscaled ecological, socioeconomic, and human behavior models to project possible futures at appropriate scales.

Scenarios to Project Climate Change, Associated Risks, and Effectiveness of Mitigation Policies

Over the past decade, the climate change research community developed a scenario framework that instead of providing one set of internally consistent and plausible visions of the future, provided a tool kit that includes GHG emission pathways (RCPs, or Representative Concentration Pathways; published in 2011), socioeconomic development pathways (SSPs, or Shared Socioeconomic Pathways; published in 2017), and possible policies. The RCPs are input into climate models for projections that do not correspond to a specific societal pathway. The SSPs are alternative societal futures, including inequities, that are as independent as possible from climate change. This framework design provides a flexible approach to addressing a range of questions. Examples include the following: Given a particular emission pathway, to what extent could development choices affect the range of possible risks? Given a particular development pathway, to what extent could different emission pathways and associated climate-related changes affect the range of possible risks? Figure 5.2 illustrates

---

[1] Use-inspired research entails engagement with a wide spectrum of users, so as to produce research that informs decision making and leverage the value of discovery-driven research (Clark et al., 2016; Stokes, 1997).

Crosscutting Research and Data Priorities

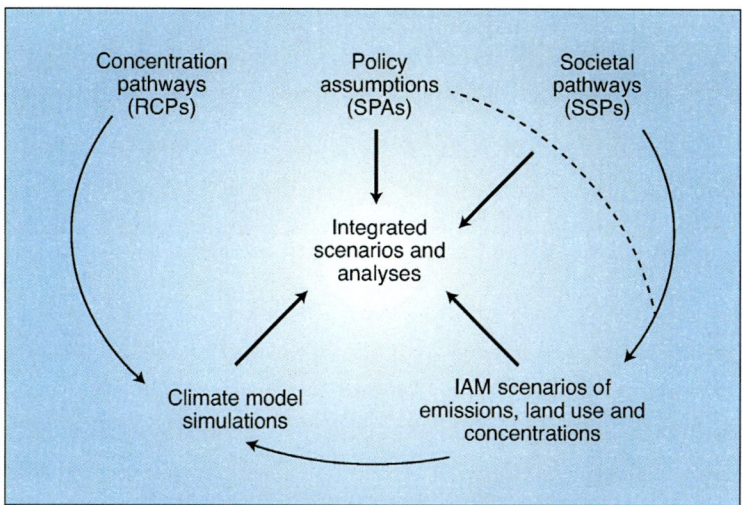

FIGURE 5.2. A scenario framework for integrating studies combining future climate outcomes, societal conditions, and policy options. This figure shows the elements of the framework and how they inform integrated scenarios and analyses. Notes: SPA = Shared Policy Assumptions; IAM = Integrated Assessment Model. Source: O'Neill et al., 2020.

the scenario framework and process for integrating studies combining future climate outcomes, societal conditions, and policy options.

The RCP-SSP framework has been widely adopted across research communities, with about 1,600 publications related to climate change drivers, risks, and response options (O'Neill et al., 2020). To date, the SSP narratives have been designed to support qualitative and quantitative extensions by region (e.g., Europe, New Zealand), by selected cities (e.g., Tokyo), and by sector (e.g., health, energy, agriculture, forestry, fisheries).

Developing narrative physical climate storylines of low-probability, high-consequence climate extremes could further understanding of complex interactions among the physical, ecological, economic, and societal aspects of extreme or compound events and could be used to explore uncertainties (Shepherd et al., 2018). These storylines should encompass a range of conditions including gradual changes in climate as well as possible extremes and tipping points.

Credible, reproducible, and consistent methods for the use of the SSPs across scales are needed to explore new questions (O'Neill et al., 2020). A more diverse set of global SSPs could facilitate exploration of a broader set of boundary conditions for multiscale analyses. This could include development of SSP variants or the mapping of other scenarios or scenario families to the SSP framework. Developing sanctioned regional scenarios would facilitate consistency across different research endeavors and organizations, such as produced for the Fourth National Climate Assessment.

Adaptation scenarios are needed that describe the transitions by which adaptation outcomes could be achieved. Projecting future resilience would be improved by incorporating variables describing strength of governance and political institutions, health care access, social protection, and other factors.

The RCP-SSP framework does not currently incorporate or address the potential for solar geoengineering research and deployment and its climatic, ecological, socioeconomic, or geopolitical implications. A challenge for including solar geoengineering research is that scenarios are static, but the feedbacks from solar geoengineering would be dynamic on relatively short timescales.

Participatory Scenario Exercises

Participatory scenario development engages decision makers and the public and incorporates normative elements that are part of human decision making (including values, beliefs, norms, and existing priorities) into a process that acknowledges future uncertainty. Using scenario planning events, stakeholders can investigate interactions, synergies, and trade-offs among goals and strategies, rather than focusing on a single outcome (Carpenter et al., 2015; Iwaniec et al., 2020; Jordan et al., 2018; Sterling et al., 2019). These scenario development processes can help identify resilience measures and motivate change at varying levels of government and with the public. The inclusion of the national security community in these participatory exercises would also benefit decision makers who could gain insight into defense and intelligence agencies' critical expertise. Once decision makers understand possible impacts, they can identify factors needed for policy change, including innovative approaches to reduce environmental hazards, increase resilience, and address inequities in the impacts of climate change and climate policy. Further research may be used to explore the implications of these interactions (Thompson et al., 2020). For example, scenarios focused on banking water versus those promoting urban greening for central Arizona in 2060 showed clear differences in the burden of extreme heat (Iwaniec et al., 2020).

Incorporating tabletop and functional planning exercises (e.g., stress testing and war games) may be effective in ground-truthing applied research findings into planning, policy, and program options considered by decision makers and the public at local, regional, state, and tribal levels. For example, back-casting approaches (i.e., starting with a desirable future and working backward rather than starting with projections and assessing how to make future outcomes more desirable) identify desirable aspects of the future; determine obstacles, including climate change, for achieving the desired goals; and identify strategies to achieve the desirable outcomes (e.g., Nikolakis, 2020). Providing web-based and/or other opportunities for broader public participation in scenario exercises should be considered to help enhance general understanding of possible outcomes and to support difficult policy decisions.

Scenarios can make climate-impact science more usable and help speed adaptation action by decision makers. Complex environmental-threat information and computational model outputs become accessible and usable for decision makers who rely on scientific and technical advisers for leading-edge guidance. When model outputs are used in the context of participatory scenarios, they are one of several inputs to the process of developing visions for a specified time and place. Rather than being seen as a future prediction that is locked in, participants can explore changes in policy, infrastructure, or distributions of natural ecosystem and built elements that might lessen or increase the impact of an environmental threat projected by the model.

The use of socioeconomic and climate storylines as elements of scenario-based problem-solving has been effective in many environmental remediation and climate action programs (Baker et al., 2004; Carpenter et al., 2015; Iwaniec et al., 2020; Shepherd et al., 2018; Thompson et al., 2020). Such programs are effective in communicating risk via narrative, hands-on, and visual depictions of actual and potential impacts to regional stakeholders. One example of a successful program model is the Adapting to Rising Tides[2] initiative in the Northern California San Francisco Bay Area, which uses both community narrative and sophisticated risk graphics to inform environmental policy and community safety improvements with state-of-the-practice scenario planning. Another example is the Dutch Dialogues, held in Charleston, South Carolina (Dutch Dialogues Charleston Team, 2019). Charleston is facing an existential crisis with tidal and storm flooding, much of it related to climate-induced sea level rise. The city is holding public discussions and engaging in planning with international and domestic flood-water management experts and local community members and leaders to develop and implement strategies to ensure that the city remains livable as climate change progresses and flooding frequency and severity increase.

---

[2] The San Francisco Bay Conservation and Development Commission Adapting to Rising Tides program is focused on helping shoreline communities in the San Francisco Bay area, spanning 10 California counties, to plan for sea level rise and other climate impacts. See http://www.adaptingtorisingtides.org.

Importance of Scale

Accelerating the exchange of technical knowledge between USGCRP agencies and decision makers in regions and communities most at risk should be a priority. Scientific findings can be made more readily actionable at decision-relevant scales when this information exchange is informed by the urgent challenges faced by local or regional entities. For example, urban practitioners are calling for downscaled climate information to understand the likelihood of future extreme events, so they can integrate their understanding of the distribution of vulnerable populations and plan for implementing defensive strategies. This is particularly important because there is increased flood exposure due to precipitation extremes and population growth in the United States (Swain et al., 2020). Coastal states need information about sea level rise and storm frequency and intensity. They also need socioeconomic projections such as population growth, property values, and availability of insurance in order to plan for coastal realignment or fortification. Municipalities dependent on water supplies from distant watersheds need information on future climate, hydrology, and ecology of those watersheds, plus projections of population and business growth in their municipality, to wisely manage future water resources. And in all these analyses, vulnerable populations and equity dimensions require special attention.

DIVERSITY, EQUITY, AND INCLUSION IN GLOBAL CHANGE RESEARCH

A significant body of research demonstrates that the risks associated with climate change are not distributed equitably across sectors, regions, or populations. In particular, racial and ethnic minorities, low-income households, and remote communities are likely to be disproportionately adversely affected by a changing climate (Dietz et al., 2020; USGCRP, 2018). Nevertheless, the U.S. public often underestimates the degree to which environmental risks are a concern, which is often very high (Pearson et al., 2018). Accordingly, local, state, and national efforts regarding climate risk management are increasingly focused on how the risks and benefits of policy interventions such as mitigation and adaptation are distributed, so that policies can be designed in a manner that is effective and fair.

Developing the evidence base to support such decision making is best facilitated by broadening participation within USGCRP science agencies and grantees, so that those populations most at-risk from climate change are represented among those conducting global change research. For example, in 2017 "Earth scientists, geologists, and oceanographers" was one of the least diverse occupation categories in the sciences, with Blacks and Hispanics comprising just 1.5 percent and 3 percent of the total despite representing 12 percent and 15 percent, respectively, of the U.S. population (NSB, 2019). This lack of inclusion undermines the capacity of the U.S. scientific enterprise to generate insights that are credible, relevant, and legitimate to diverse audiences (Cash et al., 2003). In addition, enhancing opportunities for stakeholder participation and community engagement in the sciences through transdisciplinary and community participatory research can enhance the impact and broaden the application of the science supported by USGCRP member agencies.

The committee suggests that USGCRP give a high priority to concrete efforts to increase diversity in climate science across the broad range of scientific fields and institutions involved. A first step would be to better understand the current state of, and trends in, diversity among individuals involved in research across USGCRP member agencies and, in particular, the extent to which those individuals are representative of the communities considered at greatest risk from climate change and climate policies (Avallone et al., 2013; Behl et al., 2017; Mattheis et al., 2019; Murillo et al., 2008; Popp et al., 2019). In addition, a systematic examination of the obstacles to greater diversity and an evaluation of programs and policies that have and have not been effective in increasing diversity could provide the basis for near-term action (Gay-Antaki and Liverman, 2018; Tucker et al., 2009).

In the longer term, ongoing engagement via deliberation and consultation with underrepresented communities can help build trust and engagement. The role of state, regional, and local partners can be amplified in order for research and academic experts to best frame improved decision-support findings and recommendations. Environmental and social justice organizations may also offer innovative approaches for the scientific community to apply in forming effective and productive ways to integrate heretofore underrepresented communities of influence. The committee suggests that USGCRP plans routinely and directly address actions toward the ends of building greater understanding of equity and climate justice and increasing diversity in the global change research community.

MAINTENANCE AND IMPROVEMENT OF DATA AND ANALYSIS FACILITIES

Progress in research on the risks and topics discussed in earlier chapters, as well as the previous sections on extremes, tipping points, and scenarios, calls for the design and implementation of augmented analysis frameworks that can more adequately represent these intersecting realms. Global change research to inform policies for the coming decades calls for an expanded USGCRP program on the development and management of needed data sets.

Special attention should be paid to developing and making available the social science data needed to support the security challenges of the coming decade. The culture of and mechanisms for data sharing are strong in the social sciences. The social sciences have had considerable experience designing and implementing long-term data collection efforts—some of the longest such projects are now in their seventh decade of coordinated efforts. What has been lacking is a serious and consistent federal investment in the data needed for social science analysis in support of global change research. There is an urgent need and opportunity to foster socioeconomic observing systems (Sandifer et al., 2020; Stern et al., 2013) that would link the most significant socioeconomic data streams to identify additional needed information and how it should be collected and aggregated so as to be useful for modeling future climate-related risks. Consistent with the integrated systems and risk-based approach, there is a need for greater attention to methods and analytical tools that accurately capture and integrate immaterial and non-monetized sociocultural values as part of efforts to assess co-benefits/dis-benefits of adaptation and mitigation strategies within the context of different possible development pathways.

Special attention will be needed on methods of uncertainty analysis; their incorporation into analysis of extremes, thresholds, and tipping points; and integration of results into forms that meet the particular needs of decision makers (see Box 5.1). The effort will build on current capacities such as integrated assessment models (Weyant, 2017), their coupling with agent-based models (Moss et al., 2001), the applications of life cycle analysis (Hertwich et al., 2015; Nielsen et al., 2020), and linking social science insights into modeling (Dietz et al., 2020; Nielsen et al., 2020). Maximum use needs to be made of evolving data acquisition technologies including remote sensing, exascale computing, and artificial intelligence (Reed and Dongarra, 2015).

In addition to social science data management and analysis, analytical tools and approaches supportive of the complex coupled system dynamics also need to be considered (Hallegatte et al., 2012; Ranger and Niehörster, 2012; Roelich and Giesekam, 2019). Econometric methods that analyze large data sets can also be extremely useful in providing insight into the socioeconomic impacts of climate change (Hsiang, 2016), and other big data methods such as machine and statistical learning can improve predictions (e.g., Hastie et al., 2017; Huntingford et al., 2019; Reichstein et al., 2019).

Important aspects of the analyses of natural science data have been supported primarily by nongovernment consortia. For example, The Global Carbon Project falls under the umbrella of nongovernmental organizations devoted to environmental

> **BOX 5.1**
> **Integrated Uncertainty Analysis**
>
> Uncertainty characterization and quantification is relatively well developed with regard to climate projections and with some impact models, but is less well developed for other aspects of understanding the magnitude and pattern of the risks of and responses to global change. Climate related uncertainties such as the internal variability of the climate system, the extent of emissions over coming decades, the effect of different representations of the climate system (e.g., different climate models), and how the climate systems will respond to additional radiative forcing are now well documented. However, the combined effect of these uncertainties with, for example, uncertainties regarding downscaling and internal variability are less well explored (Leduc et al., 2019).
>
> Uncertainties also exist in impact models, such as those for food or water security. Examination of the consequences of uncertainties for robustness of projections has advanced (e.g., Asseng et al., 2013), such as for combined uncertainties in crop models, economic development, and climate projections (e.g., Nelson et al., 2014). Characterizations of uncertainties becomes ever more complex as the research enterprise moves toward more completely integrated systems analyses, delineating the interactions among, for example, water, food, and health issues. What is needed is an understanding of which uncertainties have the largest impacts on projections and which could be reduced with further observation and research.
>
> No longer is filling out the linear "uncertainty cascade" adequate for representing uncertainty; uncertainty needs to be integrated across climate change research, feedbacks, and applications. For example, there is limited research documenting uncertainties in population vulnerabilities, including future uncertainties in population growth, economic security, etc. Characterizing integrated uncertainties would contribute to how the risks discussed in this report could be most effectively managed (Marchau et al., 2019).
>
> There are deep uncertainties inherent in risk management. The uncertainties surround climate science—when and where will the changes in climate occur and how large will they be? How will the natural and human systems react and adapt to these changes? In terms of scale, how is it possible to accurately project at the local or regional level where adaptation decisions are made?

research. It operates thanks to hundreds of scientists who volunteer their time and efforts to contribute to the organization and analysis of data underpinning the annual global carbon budget—the balance between its sources and sinks at a global scale. The effort of developing the global carbon budget provides a check on the estimates of its components. USGCRP should consider how to strategically leverage these and other nongovernmental efforts, as well as participate and support however possible, to facilitate long-term records for the annual global carbon budget and other key global change information data sets.

CROSSCUTTING PRIORITIES TO ADVANCE INTEGRATED SYSTEMS-BASED RISK MANAGEMENT

A number of priorities crosscut the necessary work to support the domains of risk management described previously in Chapter 4. In particular, the committee highlighted five such priorities:

1. Understanding of extreme events and tipping points, including cascading tipping points, and related feedbacks to the climate system

2. Improved simulation of local and regional-scale climate including uncertainty characterization, which can inform mitigation and adaptation responses.

3. Pursuit of a scenarios-based approach to project climate change, associated risks, and effectiveness of mitigation policies, and to increase the effectiveness of risk management through highly inclusive, stakeholder-driven processes.

4. Increased attention to uneven distribution of costs, risks and benefits of climate change and responses to it, and increased diversity in the scientific community and in the communities with whom the USGCRP engages.

5. Augmentation of existing analysis frameworks and supporting data sets to more adequately represent the many system interactions and yield results in forms that meet the needs of decision makers and the people they represent. Continued investments in research and technology, such as exascale computing, will lead to advancements that may alter the priorities of the USGCRP research agenda over the next decade.

These crosscutting topics provide opportunities for USGCRP to advance integrating efforts and cross-disciplinary research in creative ways, including, for example, through scenario-based activities (see Box 5.2).

**Box 5.2
Opportunity for Creative, Cutting-Edge Integration in Using Scenarios**

A scenarios-based approach to prioritizing research recognizes that there are unique and necessary contributions from social, physical, ecological, technological researchers *and practitioners*—i.e., those in the "trenches" seeking to implement solutions to global change threats. Rather than seeing this challenge as compartmentalized among science and engineering disciplines, this approach elevates the value of an integrated Earth system science. In this view, the "cutting edge" is redefined as those questions or approaches that are integrative from the start. For science, it means that the best questions are those asked and answered by interdisciplinary teams. For practice, the best ideas are coproduced with teams that engage the public (e.g., transdisciplinary).

The federal agencies represented in USGCRP are positioned to contribute to this transdisciplinary approach to solving the challenges of global change through creative, cutting-edge integration. Each agency can bring to the table multiple dimensions of their work, including social, physical, ecological scientists and engineers, representing the full suite of global change research being carried out. USGCRP is an ideal forum for this integration.

Recent efforts within USGCRP agencies, such as the convergence initiative at the National Science Foundation, are important steps toward achieving the transdisciplinary integration needed to address the global change challenges of the 21st century. Multiagency collaborative efforts will be needed to deal with these challenges. USGCRP has an opportunity to explore creative ways to maximize these transdisciplinary collaborative efforts.

CHAPTER SIX

Next Steps for Shifting the USGCRP Paradigm

Climate change is a grand challenge for society in the 21st century. The continued accumulation of greenhouse gases in the atmosphere and the growing impact of climate change on the lives of the American people increase the urgency of implementing effective, science-based policies to limit climate change and to manage its consequences. Similarly, policies are needed to address other critical global environmental changes, such as land use, biodiversity loss, and eutrophication of Earth's ecosystem with nitrogen. To produce the scientific understanding needed to inform such policies, the nation requires a U.S. Global Change Research Program (USGCRP or the Program) commensurate with the scope, scale, and urgency of these challenges, and with its mandates described in the Global Change Research Act of 1990 (GCRA): "a comprehensive and integrated United States research program which will assist the Nation and the world to understand, assess, predict, and respond to human-induced and natural processes of global change."

As input to USGCRP's process to develop its new strategic plan, the committee recommends that the Program apply an integrated risk-framing approach to identify research priorities. This approach would provide insights to avoid the worst potential consequences of urgent risks to human and natural systems from current and future climate change. This approach demands a paradigm shift that reorients research priorities from a historic focus on individual aspects of the natural science and the social and behavioral science to furthering understanding of coupled human-natural systems important to society, including food, water, health, transportation and infrastructure, energy systems, the economy, and national and international strategic interests (discussed in Chapter 2). The sustained security of these systems in a changing climate will be a major consideration over the coming decade.

The committee calls for USGCRP to prioritize research to manage these threats through mitigation, adaptation, and strategies that combine them (discussed in Chapters 3 and 4). An integrated systems-based approach is essential for understanding the complex consequences of concurrent mitigation and adaptation actions, and the interactions between mitigation and adaptation strategies. Such a comprehensive perspective of risk management will facilitate how the Program addresses emerging challenges posed by global change, including the co-benefits of mitigation action and the synergistic and/or antagonistic results of multiple adaptation strategies, in ways that will be useful to decision makers at multiple levels of society. Solar geoengineering, also discussed in this report, is another potential strategy that requires further research to better understand both its technical and social feasibility, characterize the potential impacts and risks of these approaches, and consider how such measures could interact with mitigation and adaptation.

This future research agenda will require the continued work of USGCRP and mission agencies to address critical gaps in fundamental understanding within natural and social sciences. As part of these efforts, the committee recommends expanded research in five crosscutting areas: (1) increasing understanding of extremes, thresholds, and tipping points; (2) projecting regional- and local-scale climate and vulnerabilities; (3) refining a scenarios-based approach to project climate change, associated risks, and effectiveness of mitigation and adaptation policies; (4) addressing multiple dimensions of equity and social justice; and (5) augmenting existing analysis frameworks and supporting data (discussed in Chapter 5).

More generally, USGCRP needs to elevate the status and support for social sciences within its portfolio to ensure that the research program yields useful information to support decisions for effectively managing shocks and stresses that can cascade through communities and states (Chapter 3). It is clear to the committee that without knowledge gained from strong social science research and the application of this knowledge to the climate change challenge, the essential societal buy-in for necessary mitigation and adaptation actions is unlikely to happen. This research needs to focus on the most vulnerable individuals, communities, and regions and promote equitable approaches and solutions.

The committee's vision for a strategic plan guided by the recommendations outlined above is both ambitious and necessary for USGCRP. The need for more integrated research, risk-management support, and advances in the identified crosscutting areas have long been recognized within the Program but not yet realized. Existing challenges facing the Program include (1) constraints on top-line budgets that have resulted in an assumption that changes must be dealt with in a "zero-sum" game; (2) the dominance of natural science-focused agencies within the Program; and (3) the emphasis within the National Science Foundation (NSF)—the agency that may be most suited to social science and cross-disciplinary research—on fundamental rather than a use-inspired combination of fundamental and applied research.

The committee's vision for USGCRP to more fully meet the mandate of the Global Change Research Act in the coming decade, given the significant climate change impacts happening today and projected to increase in the future, will require a significant expansion in scope and funding. USGCRP agencies need to maintain critical Earth system research while also providing more resources for essential social science research and research needed to couple natural and social sciences to address priority research gaps.

> **RECOMMENDATION: To accompany the shift in USGCRP paradigm, the program should explore organizational and operational changes to enhance the relevance and effectiveness of its work.**

ORGANIZATIONAL AND OPERATIONAL CHANGES

The committee provides a set of suggested organizational changes, as well as operational recommendations and proposed actions, in support of the recommendation above.

Suggested Organizational Changes

In the committee's judgment, a set of three organizational changes that fall into the category of "enlarging the tent" have the potential for adding both intellectual and financial resources in support of USGCRP's efforts to adopt the integrated systems-based approach to risk management. The first of these organizational changes focuses on federal agencies that are already part of USGCRP. These agencies need to update their level of engagement in the Program by increasing involvement of their suborganizations that bring relevant expertise and operational responsibilities to the table. For example, within the Department of Commerce, the National Institute of Standards and Technology (NIST) supports standards development for greenhouse gas emission monitoring and measurement, but other programs that develop additional relevant knowledge and standards, such as those for buildings or fire safety that are pertinent to the mitigation of or adaptation to global change impacts,[1] are not traditionally considered part of NIST's engagement with USGCRP. Another opportunity would be engaging subagencies within the National Oceanic and Atmospheric Administration (NOAA), such as the National Ocean Service and the National Marine Fisheries Service, that possess expertise required to understand and project fisheries tipping points (see Chapter 5) and inform fisheries management (NASEM, 2016a). The U.S. Department of Agriculture (USDA) includes organizations involved in preparing adaptation and mitigation strategies with respect to agriculture and rural land use,

---

[1] See https://www.gao.gov/products/GAO-17-3.

such as the Risk Management Agency, Rural Development, and the Office of the Chief Economist, which may be further, and more formally, leveraged in the Program's activities. There also may be additional opportunities within the Department of Transportation, where the Federal Highway Administration and the Federal Aviation Administration are engaged with state and local stakeholders in transportation-related resilience and mitigation efforts.[2]

The second suggested organizational change calls on USGCRP to encourage greater participation of federal mission agencies that historically have not participated in USGCRP but have relevant resources and critical expertise. As noted in the committee's 2016 report *Enhancing Participation in the U.S. Global Change Research Program*, the Program could benefit from collaborations with these mission agencies in both helping define the research agenda and in translating research into practice (NASEM, 2016a). Examples include the Department of Homeland Security and its components, such as the Federal Emergency Management Agency (FEMA), the Department of Housing and Urban Development (HUD), and the Department of Labor (NASEM, 2016a). Additionally, greater participation of agencies with international perspectives in USGCRP, such as the Fogarty International Center within the Department of Health and Human Services, National Intelligence Council, and U.S. Agency for International Development (USAID) can help to foster efforts to address global health and economic security in the context of global change (NASEM, 2016a). USGCRP would also benefit from engaging with intelligence agencies and offices that hold critical expertise, such as the Office of the Director of National Intelligence, who has examined how climate change and environmental degradation threaten U.S. national security (Kiemel, 2019; NIC, 2017).

The third suggested organizational change involves developing public-private partnerships in support of climate change research and science-to-action activities. The concept of public-private partnerships is not new for USGCRP. It was a core concept in Chapter 30 of the *Third National Climate Assessment* titled "Sustained Assessment: A New Vision for Future U.S. Assessments" (Hall et al., 2014). Besides adding intellectual and financial resources, public-private partnerships can increase the engagement of Americans in climate change—its causes, its impacts, and its solutions.

Recommended Operational Changes

The committee believes instituting four operational changes could help USGCRP reorient its approach to global change research. Throughout the report, the committee

---

[2] See, for example: https://www.globalchange.gov/agency/department-transportation; https://www.globalchange.gov/agency/department-agriculture.

has underscored the importance of considering equity and social justice as part of the climate change challenge. USGCRP has a special role to play in this critical issue. The committee strongly encourages USGCRP to champion diversity among the participants in global change science. An initial step would be an effort to better understand the current state of, and trends in, diversity among individuals involved in research across USGCRP member agencies and, in particular, the extent to which those individuals are representative of the communities considered at greatest risk from climate change and climate policies. In the longer term, ongoing engagement via deliberation and consultation between scientists and underrepresented communities would help build trust and engagement. This trust would perhaps be more easily gained if some of the scientists in the dialogue were themselves members of the underrepresented communities.

The second change focuses on coproduction, an approach used by several USGCRP agencies for more than a decade. For example, NOAA's Regional Integrated Sciences and Assessments (RISA) program and the U.S. Geological Survey's (USGS's) Climate Adaptation Science Centers (CACSs) have regularly engaged stakeholders in climate change research design and execution. As discussed in Chapter 3, research priorities need to put user needs at the forefront to ensure the research is useful, usable, and used. Further, users who are involved in setting, implementing, and communicating a research agenda are more likely to embrace its findings and information products. It would be helpful for USGCRP to develop a mechanism for evaluating the efficacy of coproduction efforts and guide the development of a list of best practices and design principles.

It is also critical that the next strategic plan outline the process through which participating agencies coordinate and adjust their individual program plans to avoid duplication and fill gaps critical to meeting overall program objectives. The committee's third operational change is therefore that USGCRP advance program integration and accountability by increasing transparency of the management structure and criteria for setting priorities, sequencing investments, and guiding development of an integrated program across the individual agencies. This process should include ongoing input from user communities on a sustained basis in keeping with effective engagement processes.

The fourth operational change is to consider developing and using mechanisms to monitor and evaluate its activities, and to use lessons learned to guide the Program's action going forward. This adaptive learning approach will allow the Program to adapt to changing priorities as scientific progress is made, and global change risks are increasingly prepared for and mitigated. The adaptive learning approach will also help the Program to better understand its potential impacts and improve the usability of its information. As part of these monitoring, evaluation, and learning efforts, the committee encourages the Program to pay special attention to the diversity issues raised in this report—the recruitment of scientists from underrepresented groups, the science the program supports, and in the communities with which the Program engages.

> **RECOMMENDATION: To enhance successful implementation of an integrated risk-management approach, it is critical that the Program does the following:**
> 1. **Prioritize diversity in both the Program and USGCRP activities by greatly expanding efforts to be inclusive and representative, and prioritize justice with research that highlights consequences and opportunities for underserved communities;**
> 2. **Increase the usability and relevance of research by adopting a coproduction approach to research, recommitting to the sustained assessment process, and establishing a standing user working group or advisory mechanism as a forum for input on user needs;**
> 3. **Advance program integration and accountability by increasing transparency of the management structure and criteria for setting priorities, sequencing investments, and guiding development of an integrated program across the individual agencies; and**
> 4. **Develop an evidence-based strategy for monitoring, evaluation, and learning for the Program's activities, including the next strategic plan, with flexibility for setting priorities and activities to adapt to and incorporate learning on an ongoing basis.**

The committee recognizes the ambitious scope of the recommended re-orientation of USGCRP's next strategic plan, given the challenges to expanding the Program. A set of potential actions are provided by the committee in Box 6.1, that may help the Program ensure an effective implementation of its new strategic plan.

FINAL THOUGHTS

The COVID-19 pandemic demonstrates the need to envision and plan for multiple, often simultaneous, and multilevel disruptions to human, physical, and ecological systems. The pandemic also provides a vivid reminder that science-based challenges should be managed using science-based policies. Specifically, such preparation for

> **BOX 6.1**
> **Actions to Help the Program Ensure an Effective Implementation of Its New Strategic Plan**
>
> - Identifying and cultivating scientific and administrative champions among the participating agencies and members of the Executive branch;
> - Articulating clear linkages to past research to build on;
> - Ensuring priorities can be well described to nonexperts;
> - Leveraging opportunities for ownership of priorities among the agency leads and/or Subcommittee on Global Change Research principals;
> - Connecting the priorities of the USGCRP strategy to existing and emerging agency priorities;
> - Developing well-defined metrics of success to accompany the priorities and progress in those areas;
> - Securing, or otherwise tapping into, adequate assets including hardware, software, and human resources; and
> - Envisioning mechanisms by which the strategy can be scaled-up (or -down, if appropriate).

multiple cascading risks requires interdisciplinary science more than ever, including the full range of disciplines across natural and social sciences.

In the future, the nation cannot afford for the scope of the Program to be based on historical budget constraints, the traditional ways that participating agencies determine their engagement in USGCRP activities, or the current identification of agencies that formally participate in USGCRP. The Global Change Research Act (GCRA) of 1990 provides the flexibility for USGCRP to include the agency participation necessary to meet the nation's needs for useful information. The GCRA also mandates that USGCRP provide readily usable information to guide effective strategies to mitigate and adapt to the effects of global change. An integrated systems-based risk-management approach as proposed by the committee in this report would enable USGCRP to meet this mandate.

Finally, the committee urges USGCRP to be bold in crafting its new strategic plan. This plan will be developed at a time when the nation is facing multiple interconnected challenges beyond climate change—COVID-19, a struggling economy, and longstanding issues related to equity and social justice. USGCRP has the opportunity to put forward a strategic plan that explains how global change research, particularly climate change research, contributes to the knowledge set needed to address these multiple interrelated challenges, and ultimately prepare society to create a more resilient future.

References

Aerts, J. C. J. H., W. J. Wouter Botzen, K. Emanuel, N. Lin, H. de Moel, and E. O. Michel-Kerjan. 2014. Evaluating flood resilience strategies for coastal megacities. *Science* 344(6183):473-475. https://doi.org/10.1126/science.1248222.

Allen, T., K. A. Murray, C. Zambrana-Torrelio, S. S. Morse, C. Rondinini, M. Di Marco, N. Breit, K. J. Olival, and P. Daszak. 2017. Global hotspots and correlates of emerging zoonotic diseases. *Nature Communications* 8(1):1124. https://doi.org/10.1038/s41467-017-00923-8.

Asseng, S., F. Ewert, C. Rosenzweig, J. W. Jones, J. L. Hatfield, A. C. Ruane, K. J. Boote, P. J. Thorburn, R. P. Rötter, D. Cammarano, N. Brisson, B. Basso, P. Martre, P. K. Aggarwal, C. Angulo, P. Bertuzzi, C. Biernath, A. J. Challinor, J. Doltra, S. Gayler, R. Goldberg, R. Grant, L. Heng, J. Hooker, L. A. Hunt, J. Ingwersen, R. C. Izaurralde, K. C. Kersebaum, C. Müller, S. Naresh Kumar, C. Nendel, G. O'Leary, J. E. Olesen, T. M. Osborne, T. Palosuo, E. Priesack, D. Ripoche, M. A. Semenov, I. Shcherbak, P. Steduto, C. Stöckle, P. Stratonovitch, T. Streck, I. Supit, F. Tao, M. Travasso, K. Waha, D. Wallach, J. W. White, J. R. Williams, and J. Wolf. 2013. Uncertainty in simulating wheat yields under climate change. *Nature Climate Change* 3(9):827-832. https://doi.org/10.1038/nclimate1916.

Avallone, L. M., A. G. Hallar, H. Thiry, and L. M. Edwards. 2013. Supporting the retention and advancement of women in the atmospheric sciences: What women are saying. *Bulletin of the American Meteorological Society* 94(9):1313-1316. https://doi.org/10.1175/BAMS-D-12-00078.1.

Baker, J. P., D. W. Hulse, S. V. Gregory, D. White, J. Van Sickle, P. A. Berger, D. Dole, and N. H. Schumaker. 2004. Alternative futures for the Willamette River Basin, Oregon. *Ecological Applications* 14(2):313-324. https://doi.org/10.1890/02-5011.

Bakker, K. 2012. Water Security: Research Challenges and Opportunities. *Science* 337(6097):914-915. https://doi.org/10.1126/science.1226337.

Beach, R. H., T. B. Sulser, A. Crimmins, N. Cenacchi, J. Cole, N. K. Fukagawa, D. Mason-D'Croz, S. Myers, M. C. Sarofim, M. Smith, and L. H. Ziska. 2019. Combining the effects of increased atmospheric carbon dioxide on protein, iron, and zinc availability and projected climate change on global diets: A modelling study. *The Lancet Planetary Health* 3(7):e307-e317. https://doi.org/10.1016/S2542-5196(19)30094-4.

Behl, M., L. Merner, and R. Pandya. 2017. Diversity at AMS: Insights from the AMS Membership Survey. *Bulletin of the American Meteorological Society* 98(9):1980.

Bidwell, D., T. Dietz, and D. Scavia. 2013. Fostering knowledge networks for climate adaptation. *Nature Climate Change* 3(7):610-611. https://doi.org/10.1038/nclimate1931.

Bleemer, Z., and W. van der Klaauw. 2019. Long-run net distributionary effects of federal disaster insurance: The case of Hurricane Katrina. *Journal of Urban Economics* 110:70-88. https://doi.org/10.1016/j.jue.2019.01.005.

Bonaccorsi, G., F. Pierri, M. Cinelli, A. Flori, A. Galeazzi, F. Porcelli, A. L. Schmidt, C. M. Valensise, A. Scala, W. Quattrociocchi, and F. Pammolli. 2020. Economic and social consequences of human mobility restrictions under COVID-19. *Proceedings of the National Academy of Sciences* 117(27):15530-15535. https://doi.org/10.1073/pnas.2007658117.

Brown, M. E., J. M. Antle, P. Backlund, E. R. Carr, W. E. Easterling, M. K. Walsh, C. Ammann, W. Attavanich, C. B. Barrett, M. F. Bellemare, V. Dancheck, C. Funk, K. Grace, J. S. I. Ingram, H. Jiang, H. Maletta, T. Mata, A. Murray, M. Ngugi, D. Ojima, B. O'Neill, and C. Tebaldi. 2015. Climate Change, Global Food Security, and the U.S. Food System. http://www.usda.gov/oce/climate_change/FoodSecurity2015Assessment/FullAssessment.pdf.

Byers, E., M. Gidden, D. Leclère, J. Balkovic, P. Burek, K. Ebi, P. Greve, D. Grey, P. Havlik, A. Hillers, N. Johnson, T. Kahil, V. Krey, S. Langan, N. Nakicenovic, R. Novak, M. Obersteiner, S. Pachauri, A. Palazzo, S. Parkinson, N. D. Rao, J. Rogelj, Y. Satoh, Y. Wada, B. Willaarts, and K. Riahi. 2018. Global exposure and vulnerability to multi-sector development and climate change hotspots. *Environmental Research Letters* 13(5):055012. https://doi.org/10.1088/1748-9326/aabf45.

Campbell, R. J., C. E. Clark, and D. A. Austin. 2017. *Repair or Rebuild: Options for Electric Power in Puerto Rico.* R45023. Washington, DC: Congressional Research Service. https://fas.org/sgp/crs/row/R45023.pdf.

Caniglia, G., C. Luederitz, T. von Wirth, I. Fazey, B. Martín-López, K. Hondrila, A. König, H. von Wehrden, N. A. Schäpke, M. D. Laubichler, and D. J. Lang. 2021. A pluralistic and integrated approach to action-oriented knowledge for sustainability. *Nature Sustainability* 4(2):93-100. https://doi.org/10.1038/s41893-020-00616-z.

Carpenter, S. R., E. G. Booth, S. Gillon, C. J. Kucharik, S. Loheide, A. S. Mase, M. Motew, J. Qiu, A. R. Rissman, J. Seifert, E. Soylu, M. Turner, and C. B. Wardropper. 2015. Plausible futures of a social-ecological system: Yahara watershed, Wisconsin, USA. *Ecology and Society* 20(2). https://doi.org/10.5751/ES-07433-200210.

Cash, D. W., W. C. Clark, F. Alcock, N. M. Dickson, N. Eckley, D. H. Guston, J. Jäger, and R. B. Mitchell. 2003. Knowledge systems for sustainable development. *Proceedings of the National Academy of Sciences* 100(14):8086-8091. https://doi.org/10.1073/pnas.1231332100.

Center for Climate and Security. 2019. *A Climate Security Plan for America*. Washington, DC: The Climate and Security Advisory Group, Chaired by the Center for Climate and Security in partnership with George Washington University's Elliott School of International Affairs. https://climateandsecurity.org/wp-content/uploads/2019/09/a-climate-security-plan-for-america_2019_9_24-1.pdf.

CES (Committee on Earth Sciences). 1989. *Our Changing Planet: A US Strategy for Global Change Research*. Washington, DC: Office of Science and Technology Policy.

Chetty, R., N. Hendren, M. R. Jones, and S. R. Porter. 2018. *Race and Economic Opportunity in the United States: An Intergenerational Perspective*. Cambridge, MA: National Bureau of Economic Research. https://www.nber.org/papers/w24441.

Clark, W. C., L. van Kerkhoff, L. Lebel, and G. C. Gallopin. 2016. Crafting usable knowledge for sustainable development. *Proceedings of the National Academy of Sciences* 113(17):4570-4578. https://doi.org/10.1073/pnas.1601266113.

Clarke, L., L. Nichols, R. Vallario, M. Hejazi, J. Horing, A. C. Janetos, K. Mach, M. Mastrandrea, M. Orr, B. L. Preston, R. Reed, R. D. Sands, and D. D. White. 2018. Sector interactions, multiple stressors, and complex systems. In *Impacts, Risks, and Adaptation in the United States: Fourth National Climate Assessment, Volume II*. D. R. Reidmiller, C. W. Avery, D. R. Easterling, K. E. Kunkel, K. L. M. Lewis, T. K. Maycock and B. C. Stewart, eds. Washington, DC: U.S. Global Change Research Program.

Clément, V., H. Rey-Valette, and B. Rulleau. 2015. Perceptions on equity and responsibility in coastal zone policies. *Ecological Economics* 119:284-291. https://doi.org/10.1016/j.ecolecon.2015.09.005.

CNA Corporation. 2007. National Security and the Threat of Climate Change. https://www.cna.org/cna_files/pdf/national%20security%20and%20the%20threat%20of%20climate%20change.pdf.

CNA Corporation. 2017. *The Role of Water Stress in Instability and Conflict*. CRM-2017-U-016532. https://www.cna.org/CNA_files/pdf/CRM-2017-U-016532-Final.pdf.

Coats, D. R. 2019. Statement for the Record: 2019 Worldwide Threat Assessment of the U.S. Intelligence Community. https://www.odni.gov/index.php/newsroom/congressional-testimonies/item/1947-statement-for-the-record-worldwide-threat-assessment-of-the-us-intelligence-community.

Cooke, R., B. A. Wielicki, D. F. Young, and M. G. Mlynczak. 2014. Value of information for climate observing systems. *Environment Systems and Decisions* 34(1):98-109. https://doi.org/10.1007/s10669-013-9451-8.

Dannenberg, A. L., H. Frumkin, J. J. Hess, and K. L. Ebi. 2019. Managed retreat as a strategy for climate change adaptation in small communities: Public health implications. *Climatic Change* 153(1):1-14. https://doi.org/10.1007/s10584-019-02382-0.

Daszak, P., A. A. Cunningham, and A. D. Hyatt. 2001. Anthropogenic environmental change and the emergence of infectious diseases in wildlife. *Acta Tropica* 78(2):103-116. https://doi.org/10.1016/s0001-706x(00)00179-0.

Deryugina, T., L. Kawano, and S. Levitt. 2018. The economic impact of Hurricane Katrina on its victims: Evidence from individual tax returns. *American Economic Journal: Applied Economics* 10(2):202-233. https://doi.org/10.1257/app.20160307.

Dietz, T., R. L. Shwom, and C. T. Whitley. 2020. Climate Change and Society. *Annual Review of Sociology* 46(1):135-158. https://doi.org/10.1146/annurev-soc-121919-054614.

DOE (U.S. Department of Energy). 2014. *Effect of Sea Level Rise on Energy Infrastructure in Four Major Metropolitan Areas*. Washington, DC: U.S. Department of Energy Office of Electricity Delivery and Energy Reliability. https://www.energy.gov/sites/prod/files/2014/10/f18/DOE-OE_SLR%20Public%20Report_Final%20_2014-10-10.pdf.

DOE. 2017. *Quadrennial Energy Review. Transforming the Nation's Electricity System: The Second Installment of the QER*. Washington, DC: Department of Energy, Office of Policy. https://www.energy.gov/policy/downloads/quadrennial-energy-review-second-installment.

References

DOT (U.S. Department of Transportation) and FHWA (Federal Highway Administration). 2013. Risk-Based Transportation Asset Management: Building Resilience into Ttransportation Assets, Report 5: Managing External Threats Through Risk-Based Asset Management. https://www.fhwa.dot.gov/asset/pubs/hif13018.pdf.

Dryzek, J. S., D. Nicol, S. Niemeyer, S. Pemberton, N. Curato, A. Bächtiger, P. Batterham, B. Bedsted, S. Burall, M. Burgess, G. Burgio, Y. Castelfranchi, H. Chneiweiss, G. Church, M. Crossley, J. de Vries, M. Farooqe, M. Hammond, B. He, R. Mendonça, J. Merchant, A. Middleton, J. E. J. Rasko, I. Van Hoyweghen, and A. Vergne. 2020. Global citizen deliberation on genome editing. *Science* 369(6510):1435-1437. https://doi.org/10.1126/science.abb5931.

Dutch Dialogues Charleston Team. 2019. Dutch Dialogues Charleston. https://www.historiccharleston.org/dutch-dialogues.

Ebi, K. L., L. H. Ziska, and G. W. Yohe. 2016. The shape of impacts to come: Lessons and opportunities for adaptation from uneven increases in global and regional temperatures. *Climatic Change* 139(3):341-349. https://doi.org/10.1007/s10584-016-1816-9.

Ebi, K. L., T. Hasegawa, K. Hayes, A. Monaghan, S. Paz, and P. Berry. 2018. Health risks of warming of 1.5°C, 2°C, and higher, above pre-industrial temperatures. *Environmental Research Letters* 13(6):063007. https://doi.org/10.1088/1748-9326/aac4bd.

Eriksen, S., E. L. F. Schipper, M. Scoville-Simonds, K. Vincent, H. N. Adam, N. Brooks, B. Harding, D. Khatri, L. Lenaerts, D. Liverman, M. Mills-Novoa, M. Mosberg, S. Movik, B. Muok, A. Nightingale, H. Ojha, L. Sygna, M. Taylor, C. Vogel, and J. J. West. 2021. Adaptation interventions and their effect on vulnerability in developing countries: Help, hindrance or irrelevance? *World Development* 141:105383. https://doi.org/10.1016/j.worlddev.2020.105383.

Fan, Q. I. N., and M. Davlasheridze. 2018. Economic impacts of migration and brain drain after major catastrophe: The case of Hurricane Katrina. *Climate Change Economics* 10(01):1950004. https://doi.org/10.1142/S2010007819500040.

Fetzek, S., and L. van Schaik. 2018. *Europe's Responsibility to Prepare: Managing Climate Security Risks in a Changing World*. Washington, DC: Center for Climate and Security. https://climateandsecurity.org/wp-content/uploads/2018/06/europes-responsibility-to-prepare_managing-climate-security-risks-in-a-changing-world_2018_6.pdf.

Gasteyer, S. P., J. Lai, B. Tucker, J. Carrera, and J. Moss. 2016. Basics inequality: Race and access to complete plumbing facilities in the United States. *Du Bois Review: Social Science Research on Race* 13(2):305-325. https://doi.org/10.1017/S1742058X16000242.

Gay-Antaki, M., and D. Liverman. 2018. Climate for women in climate science: Women scientists and the Intergovernmental Panel on Climate Change. *Proceedings of the National Academy of Sciences of the United States of America* 115(9):2060-2065. https://doi.org/10.1073/pnas.1710271115.

GHSI. 2019. 2019 Global Health Security Index. https://www.ghsindex.org.

Gilbert, A. Q., and B. K. Sovacool. 2016. Looking the wrong way: Bias, renewable electricity, and energy modelling in the United States. *Energy* 94:533-541. https://doi.org/10.1016/j.energy.2015.10.135.

Gilligan, J. M., and M. P. Vandenbergh. 2020. A framework for assessing the impact of private climate governance. *Energy Research & Social Science* 60:101400. https://doi.org/10.1016/j.erss.2019.101400.

Gleick, P., and C. Iceland. 2018. *Water, Security and Conflict*. Washington, DC: World Resources Institute. https://www.wri.org/publication/water-security-and-conflict.

Global Carbon Project. 2020. The Global Carbon Project. https://www.globalcarbonproject.org.

Government Accountability Office. 2014. *Climate Change Energy Infrastructure Risks and Adaptation Efforts*. GAO-14-74. Washington, DC: Government Accountability Office. https://www.gao.gov/assets/670/660558.pdf.

Gunderson, R., and T. Dietz. 2018. Deliberation and Catastrophic Risks. In *Oxford Handbook of Deliberative Democracy*. A. Bächtiger, J. Mansbridge, M. E. Warren and J. Dryzek, eds. Oxford, UK: Oxford University Press.

Guy, K. A., J. Conger, F. Femia, S. Goodman, L. Hering Sr, A. C. Hill, A. D. Jameson, R. D. Kauzlarich, R. Keys, M. D. King, R. Schoonover, J. D. B. VanDervort, S. Veeravalli, C. E. Werrell, and P. Zukunft. 2020. *A Security Threat Assessment of Global Climate Change: How Likely Warming Scenarios Indicate a Catastrophic Security Future*. Product of the National Security, Military, and Intelligence Panel on Climate Change. F. Femia and C. Werrell, eds. Washington, DC: The Center for Climate and Security, an institute of the Council on Strategic Risks.

Hacker, J. S., G. A. Huber, A. Nichols, P. Rehm, M. Schlesinger, R. Valletta, and S. Craig. 2014. The Economic Security Index: A new measure for research and policy analysis. *The Review of Income and Wealth* 60(S1):S5-S32. https://doi.org/10.1111/roiw.12053C.

Haimes, Y. Y. 2018. Risk modeling of interdependent complex systems of systems: Theory and practice. *Risk Analysis* 38(1):84-98. https://doi.org/10.1111/risa.12804.

Hall, J. A., M. Blair, J. L. Buizer, D. I. Gustafson, B. Holland, S. C. Moser, and A. M. Waple. 2014. Sustained Assessment: A New Vision for Future U.S. Assessments. In J. M. Melillo, T. C. Richmond, and G. W. Yohe, eds. *Climate Change Impacts in the United States: The Third National Climate Assessment*. Washington, DC: U.S. Global Change Research Program. https://doi.org/10.7930/J0Z31WJ2.

Hallegatte, S., A. Shah, R. Lempert, C. Brown, and S. Gill. 2012. *Investment Decision Making under Deep Uncertainty—Application to Climate Change*. Washington, DC: The World Bank.

Hamilton, L. C., R. L. Haedrich, and C. M. Duncan. 2004. Above and below the Water: Social/Ecological Transformation in Northwest Newfoundland. *Population and Environment* 25(3):195-215.

Harari, M., and E. L. Ferrara. 2018. Conflict, climate, and cells: A disaggregated analysis. *The Review of Economics and Statistics* 100(4):594-608.

Hastie, T., R. Tibshirani, and J. Friedman. 2017. *The Elements of Statistical Learning: Data Mining, Inference, and Prediction*. New York: Springer Science.

Helbing, D. 2013. Globally networked risks and how to respond. *Nature* 497(7447):51-59. https://doi.org/10.1038/nature12047.

Helliwell, J. F., H. Huang, and S. Wang. 2014. Social capital and well-being in times of crisis. *Journal of Happiness Studies* 15(1):145-162. https://doi.org/10.1007/s10902-013-9441-z.

Hepburn, C., B. O'Callaghan, N. Stern, J. Stiglitz, and D. Zenghelis. 2020. Will COVID-19 fiscal recovery packages accelerate or retard progress on climate change? *Oxford Review of Economic Policy* 36(Supplement_1):S359-S381. https://doi.org/10.1093/oxrep/graa015.

Hertwich, E. G., T. Gibon, E. A. Bouman, A. Arvesen, S. Suh, G. A. Heath, J. D. Bergesen, A. Ramirez, M. I. Vega, and L. Shi. 2015. Integrated life-cycle assessment of electricity-supply scenarios confirms global environmental benefit of low-carbon technologies. *Proceedings of the National Academy of Sciences* 112(20):6277-6282. https://doi.org/10.1073/pnas.1312753111.

Hoffman, J. S., V. Shandas, and N. Pendleton. 2020. The effects of historical housing policies on resident exposure to intra-urban heat: A study of 108 US urban areas. *Climate* 8(1):12. https://doi.org/10.3390/cli8010012.

Holland, G., and C. L. Bruyère. 2014. Recent intense hurricane response to global climate change. *Climate Dynamics* 42(3):617-627. https://doi.org/10.1007/s00382-013-1713-0.

Holm, P., and V. Winiwarter. 2017. Climate change studies and the human sciences. *Global and Planetary Change* 156:115-122. https://doi.org/10.1016/j.gloplacha.2017.05.006.

Hosseini, S. E. 2020. An outlook on the global development of renewable and sustainable energy at the time of COVID-19. *Energy Research & Social Science* 68:101633. https://doi.org/10.1016/j.erss.2020.101633.

Hsiang, S. 2016. Climate econometrics. *Annual Review of Resource Economics* 8(1):43-75. https://doi.org/10.1146/annurev-resource-100815-095343.

Huntingford, C., E. S. Jeffers, M. B. Bonsall, H. M. Christensen, T. Lees, and H. Yang. 2019. Machine learning and artificial intelligence to aid climate change research and preparedness. *Environmental Research Letters* 14(12):124007. https://doi.org/10.1088/1748-9326/ab4e55.

IEA (International Energy Agency). 2020. World Energy Outlook. Paris: International Energy Agency. https://www.iea.org/reports/world-energy-outlook-2020.

IEA. 2021. *Covid-19 impact on electricity*. Paris: International Energy Agency. https://www.iea.org/reports/covid-19-impact-on-electricity.

IMCCS (International Military Council on Climate and Security). 2020. *The World Climate and Security Report 2020*. F. Femia and C. Werrell, eds. Washington, DC: Center for Climate and Security. https://climateandsecurity.org/wp-content/uploads/2020/02/world-climate-security-report-2020_2_13.pdf.

IPCC (Intergovernmental Panel on Climate Change). 2006. *2006 IPCC Guidelines for National Greenhouse Gas Inventories*. H.S. Eggleston, L. Buendia, K. Miwa, T. Ngara, and K. Tanabe, eds. Hayama, Kanagawa, Japan: Institute for Global Environmental Strategies.

References

IPCC. 2012. *Managing the Risks of Extreme Events and Disasters to Advance Climate Change Adaptation. A Special Report of Working Groups I and II of the Intergovernmental Panel on Climate Change*. A Special Report of Working Groups I and II of the Intergovernmental Panel on Climate Change. C. B. Field, V. Barros, T. F. Stocker, D. Qin, D. J. Dokken, K. L. Ebi, M. D. Mastrandrea, K. J. Mach, G.-K. Plattner, S. K. Allen, M. Tignor, and P. M. Midgley, eds. Cambridge, UK, and New York, NY: Cambridge University Press.

IPCC. 2014. *Climate Change 2014: Impacts, Adaptation, and Vulnerability. Part A: Global and Sectoral Aspects. Contribution of Working Group II to the Fifth Assessment Report of the Intergovernmental Panel on Climate Change*. C. B. Field, V. R. Barros, D. J. Dokken, K. J. Mach, M. D. Mastrandrea, T. E. Bilir, M. Chatterjee, K. L. Ebi, Y. O. Estrada, R. C. Genova, B. Girma, E. S. Kissel, A. N. Levy, S. MacCracken, P. R. Mastrandrea, and L. L. White, eds. Cambridge, UK and New York, NY: Cambridge University Press.

IPCC. 2018. *Global Warming of 1.5°C. An IPCC Special Report on the impacts of global warming of 1.5°C above pre-industrial levels and related global greenhouse gas emission pathways, in the context of strengthening the global response to the threat of climate change, sustainable development, and efforts to eradicate poverty*. V. Masson-Delmotte, P. Zhai, H.-O. Pörtner, D. Roberts, J. Skea, P.R. Shukla, A. Pirani, W. Moufouma-Okia, C. Péan, R. Pidcock, S. Connors, J.B.R. Matthews, Y. Chen, X. Zhou, M.I. Gomis, E. Lonnoy, T. Maycock, M. Tignor, and T. Waterfield, eds. Geneva, Switzerland: World Meteorological Organization.

IPCC. 2019a. *Climate Change and Land: an IPCC special report on climate change, desertification, land degradation, sustainable land management, food security, and greenhouse gas fluxes in terrestrial ecosystems*. P.R. Shukla, J. Skea, E. Calvo Buendia, V. Masson-Delmotte, H.-O. Pörtner, D. C. Roberts, P. Zhai, R. Slade, S. Connors, R. van Diemen, M. Ferrat, E. Haughey, S. Luz, S. Neogi, M. Pathak, J. Petzold, J. Portugal Pereira, P. Vyas, E. Huntley, K. Kissick, M. Belkacemi, and J. Malley, eds. In Press.

IPCC. 2019b. *IPCC Special Report on the Ocean and Cryosphere in a Changing Climate*. H.-O. Pörtner, D. C. Roberts, V. Masson-Delmotte, P. Zhai, M. Tignor, E. Poloczanska, K. Mintenbeck, A. Alegría, M. Nicolai, A. Okem, J. Petzold, B. Rama, and N. M. Weyer, eds. Geneva, Switzerland: IPCC.

IPCC. 2019c. *2019 Refinement to the 2006 IPCC Guidelines for National Greenhouse Gas Inventories*. E. Calvo Buendia, K. Tanabe, A. Kranjc, J. Baasansuren, M. Fukuda, S. Ngarize, A. Osako, Y. Pyrozhenko, P. Shermanau, and S. Federici, eds. Geneva, Switzerland: IPCC.

Iwaniec, D. M., E. M. Cook, M. J. Davidson, M. Berbés-Blázquez, M. Georgescu, E. S. Krayenhoff, A. Middel, D. A. Sampson, and N. B. Grimm. 2020. The co-production of sustainable future scenarios. *Landscape and Urban Planning* 197:103744. https://doi.org/10.1016/j.landurbplan.2020.103744.

Jakob, M., J. C. Steckel, F. Jotzo, B. K. Sovacool, L. Cornelsen, R. Chandra, O. Edenhofer, C. Holden, A. Löschel, T. Nace, N. Robins, J. Suedekum, and J. Urpelainen. 2020. The future of coal in a carbon-constrained climate. *Nature Climate Change* 10(8):704-707. https://doi.org/10.1038/s41558-020-0866-1.

Janetos, A. C. 2020. Why is climate adaptation so important? What are the needs for additional research? *Climatic Change* 161(1):171-176. https://doi.org/10.1007/s10584-019-02651-y.

Jordan, R., S. Gray, M. Zellner, P. D. Glynn, A. Voinov, B. Hedelin, E. J. Sterling, K. Leong, L. S. Olabisi, K. Hubacek, P. Bommel, T. K. BenDor, A. J. Jetter, B. Laursen, A. Singer, P. J. Giabbanelli, N. Kolagani, L. B. Carrera, K. Jenni, and C. Prell. 2018. Twelve questions for the participatory modeling community. *Earth's Future* 6(8):1046-1057. https://doi.org/10.1029/2018EF000841.

Keisler, J. M., Z. A. Collier, E. Chu, N. Sinatra, and I. Linkov. 2014. Value of information analysis: The state of application. *Environment Systems and Decisions* 34(1):3-23. https://doi.org/10.1007/s10669-013-9439-4.

Kiemel, P. 2019. Statement for the Record for a Hearing on "The National Security Implications of Climate Change" before the Permanent Select Committee on Intelligence, US House of Representatives. https://www.dni.gov/files/ODNI/documents/2019-6-05_Statement_-_HPSCI_Climate_Change_Hearing_-_APPROVED_converted.pdf.

Kim, L., J. Marlon, M. Ballew, and K. Lacroix. 2020. *How does the American public perceive climate disasters?* Yale Program on Climate Change Communication. New Haven, CT: Yale School of the Environment. https://climatecommunication.yale.edu/publications/how-does-the-american-public-perceive-climate-disasters.

Klarevas, L., and C. P. Clarke. 2020. Is COVID-19 a National Security Emergency? The RAND Blog, August 6. https://www.rand.org/blog/2020/08/is-covid-19-a-national-security-emergency.html.

Kopp, R. E., K. Hayhoe, D. R. Easterling, T. Hall, R. Horton, K. E. Kunkel, and A. N. LeGrande. 2017. Potential Surprises: Compound Extremes and Tipping Elements. In *Climate Science Special Report: Fourth National Climate Assessment*. D. J. Wuebbles, D. W. Fahey, K. A. Hibbard, D. J. Dokken, B. C. Stewart and T. K. Maycock, eds. Washington, DC: U.S. Global Change Research Program.

Kossin, J. P., T. Hall, T. Knutson, K. E. Kunkel, R. J. Trapp, D. E. Waliser, and M. F. Wehner. 2017. Extreme Storms. In *Climate Science Special Report: Fourth National Climate Assessment*. D. J. Wuebbles, D. W. Fahey, K. A. Hibbard, D. J. Dokken, B. C. Stewart and T. K. Maycock, eds. Washington, DC: U.S. Global Change Research Program.

Leduc, M., A. Mailhot, A. Frigon, J.-L. Martel, R. Ludwig, G. B. Brietzke, M. Giguère, F. Brissette, R. Turcotte, M. Braun, and J. Scinocca. 2019. The ClimEx Project: A 50-member ensemble of climate change projections at 12-km resolution over Europe and Northeastern North America with the Canadian Regional Climate Model (CRCM5). *Journal of Applied Meteorology and Climatology* 58(4):663-693. https://doi.org/10.1175/JAMC-D-18-0021.1.

Lemos, M. C., C. J. Kirchhoff, S. E. Kalafatis, D. Scavia, and R. B. Rood. 2014. Moving climate information off the shelf: Boundary chains and the role of RISAs as adaptive organizations. *Weather, Climate, and Society* 6(2):273-285. https://doi.org/10.1175/WCAS-D-13-00044.1.

Lenton, T. M., J. Rockström, O. Gaffney, S. Rahmstorf, K. Richardson, W. Steffen, and H. J. Schellnhuber. 2019. Climate tipping points—too risky to bet against. *Nature* 575(7784):592-595. https://doi.org/10.1038/d41586-019-03595-0.

Loh, E. H., C. Zambrana-Torrelio, K. J. Olival, T. L. Bogich, C. K. Johnson, J. A. Mazet, W. Karesh, and P. Daszak. 2015. Targeting transmission pathways for emerging zoonotic disease surveillance and control. *Vector Borne Zoonotic Diseases* 15(7):432-437. https://doi.org/10.1089/vbz.2013.1563.

Loladze, I. 2014. Hidden shift of the ionome of plants exposed to elevated CO_2 depletes minerals at the base of human nutrition. *eLife* 3:e02245. https://doi.org/10.7554/eLife.02245.

Mach, K. J., C. M. Kraan, M. Hino, A. R. Siders, E. M. Johnston, and C. B. Field. 2019. Managed retreat through voluntary buyouts of flood-prone properties. *Science Advances* 5(10). https://doi.org/10.1126/sciadv.aax8995.

Maloney, M. C., and B. L. Preston. 2014. A geospatial dataset for U.S. hurricane storm surge and sea-level rise vulnerability: Development and case study applications. *Climate Risk Management* 2:26-41. https://doi.org/10.1016/j.crm.2014.02.004.

Marchau, V. A. W. J., W. E. Walker, P. J. T. M. Bloemen, and S. W. Popper, eds. 2019. *Decision Making under Deep Uncertainty: From Theory to Practice*. Cham, Switzerland: Springer.

Martinich, J., and A. Crimmins. 2019. Climate damages and adaptation potential across diverse sectors of the United States. *Nature Climate Change* 9(5):397-404. https://doi.org/10.1038/s41558-019-0444-6.

Masten, S. J., S. H. Davies, and S. P. McElmurry. 2016. Flint Water Crisis: What Happened and Why? *Journal - American Water Works Association* 108(12):22-34. https://doi.org/10.5942/jawwa.2016.108.0195.

Mattheis, A., M. Murphy, and E. Marin-Spiotta. 2019. Examining intersectionality and inclusivity in geosciences education research: A synthesis of the literature 2008–2018. *Journal of Geoscience Education* 67(4):505-517. https://doi.org/10.1080/10899995.2019.1656522.

McClymont Peace, D., and E. Myers. 2012. Community-based participatory process – Climate change and health adaptation program for Northern First Nations and Inuit in Canada. *International Journal of Circumpolar Health* 71(1):18412. https://doi.org/10.3402/ijch.v71i0.18412.

McIntosh, K., E. Ross, R. Nunn, and J. Shambaugh. 2020. *Examining the Black-White Wealth Gap*. Washington, DC: Brookings Institution. https://www.brookings.edu/blog/up-front/2020/02/27/examining-the-black-white-wealth-gap.

McLaughlin, P. 2011. Climate change, adaptation, and vulnerability: Reconceptualizing societal–environment interaction within a socially constructed adaptive landscape. *Organization & Environment* 24(3):269-291. https://doi.org/10.1177/1086026611419862.

Moallemi, E. A., J. Kwakkel, F. J. de Haan, and B. A. Bryan. 2020. Exploratory modeling for analyzing coupled human-natural systems under uncertainty. *Global Environmental Change* 65:102186. https://doi.org/10.1016/j.gloenvcha.2020.102186.

Morand, S., and B. Walther. 2020. The accelerated infectious disease risk in the Anthropocene: More outbreaks and wider global spread. *bioRxiv*. https://doi.org/10.1101/2020.04.20.049866.

Moss, R. H., S. Avery, K. Baja, M. Burkett, A. M. Chischilly, J. Dell, P. A. Fleming, K. Geil, K. Jacobs, A. Jones, K. Knowlton, J. Koh, M. C. Lemos, J. Melillo, R. Pandya, T. C. Richmond, L. Scarlett, J. Snyder, M. Stults, A. M. Waple, J. Whitehead, D. Zarrilli, B. M. Ayyub, J. Fox, A. Ganguly, L. Joppa, S. Julius, P. Kirshen, R. Kreutter, A. McGovern, R. Meyer, J. Neumann, W. Solecki, J. Smith, P. Tissot, G. Yohe, and R. Zimmerman. 2019. Evaluating knowledge to support climate action: A framework for sustained assessment. Report of an Independent Advisory Committee on Applied Climate Assessment. *Weather, Climate, and Society* 11(3):465-487. https://doi.org/10.1175/WCAS-D-18-0134.1.

Moss, S., C. Pahl-Wostl, and T. Downing. 2001. Agent-based integrated assessment modelling: The example of climate change. *Integrated Assessment* 2(1):17-30. https://doi.org/10.1023/A:1011527523183.

Murillo, S. T., R. E. Pandya, R. Y. Chu, J. A. Winkler, R. Czujko, and E. M. C. Cutrim. 2008. AMS membership survey results: An overview and longitudinal analysis of the demographics of the AMS. *Bulletin of the American Meteorological Society* 89(5):727-733.

NASEM (National Academies of Sciences, Engineering, and Medicine). 2016a. *Enhancing Participation in the U.S. Global Change Research Program*. Washington, DC: The National Academies Press.

NASEM. 2016b. *Gene Drives on the Horizon: Advancing Science, Navigating Uncertainty, and Aligning Research with Public Values*. Washington, DC: The National Academies Press.

NASEM. 2016c. *Transportation Resilience: Adaptation to Climate Change*. Washington, DC: The National Academies Press.

NASEM. 2017a. *Accomplishments of the U.S. Global Change Research Program*. Washington, DC: The National Academies Press.

NASEM. 2017b. *Human Genome Editing: Science, Ethics, and Governance*. Washington, DC: The National Academies Press.

NASEM. 2018. *Improving Characterization of Anthropogenic Methane Emissions in the United States*. Washington, DC: The National Academies Press.

NASEM. 2019a. *Negative Emissions Technologies and Reliable Sequestration: A Research Agenda*. Washington, DC: The National Academies Press.

NASEM. 2019b. *Gaseous Carbon Waste Streams Utilization: Status and Research Needs*. Washington, DC: The National Academies Press.

NASEM. 2021a. *Accelerating Decarbonization of the U.S. Energy System*. Washington, DC: The National Academies Press.

NASEM. 2021b. *Reflecting Sunlight: Recommendations for Solar Geoengineering Research and Research Governance*. Washington, DC: The National Academies Press.

Nelson, G. C., H. Valin, R. D. Sands, P. Havlík, H. Ahammad, D. Deryng, J. Elliott, S. Fujimori, T. Hasegawa, E. Heyhoe, P. Kyle, M. Von Lampe, H. Lotze-Campen, D. Mason d'Croz, H. van Meijl, D. van der Mensbrugghe, C. Müller, A. Popp, R. Robertson, S. Robinson, E. Schmid, C. Schmitz, A. Tabeau, and D. Willenbockel. 2014. Climate change effects on agriculture: Economic responses to biophysical shocks. *Proceedings of the National Academy of Sciences* 111(9):3274-3279. https://doi.org/10.1073/pnas.1222465110.

Nett, K., and L. Rüttinger. 2016. *Insurgency, Terrorism and Organised Crime in a Warming Climate. Analysing the Links Between Climate Change and Non-State Armed Groups*. Berlin: adelphi. https://www.adelphi.de/en/publication/insurgency-terrorism-and-organised-crime-warming-climate.

Neumann, B., A. T. Vafeidis, J. Zimmermann, and R. J. Nicholls. 2015. Future coastal population growth and exposure to sea-level rise and coastal flooding–A global assessment. *PLOS ONE* 10(3):e0118571. https://doi.org/10.1371/journal.pone.0118571.

NIC (National Intelligence Council). 2017. Global Trends: Paradox of Progress. https://climateandsecurity.org/wp-content/uploads/2019/03/nic_global-trends_paradox-of-progress.pdf.

Nicola, M., Z. Alsafi, C. Sohrabi, A. Kerwan, A. Al-Jabir, C. Iosifidis, M. Agha, and R. Agha. 2020. The socio-economic implications of the coronavirus pandemic (COVID-19): A review. *International Journal of Surgery (London, England)* 78:185-193. https://doi.org/10.1016/j.ijsu.2020.04.018.

Nielsen, K. S., P. C. Stern, T. Dietz, J. M. Gilligan, D. P. van Vuuren, M. J. Figueroa, C. Folke, W. Gwozdz, D. Ivanova, L. A. Reisch, M. P. Vandenbergh, K. S. Wolske, and R. Wood. 2020. Improving climate change mitigation analysis: A framework for examining feasibility. *One Earth* 3(3):325-336. https://doi.org/https://doi.org/10.1016/j.oneear.2020.08.007.

Nikolakis, W. 2020. Participatory backcasting: Building pathways towards reconciliation? *Futures* 122:102603. https://doi.org/10.1016/j.futures.2020.102603.

NOAA (National Oceanic and Atmospheric Administration). 2013. *National Coastal Population Report. Population Trends from 1970 to 2020*. Silver Spring, MD: NOAA. https://aamboceanservice.blob.core.windows.net/oceanservice-prod/facts/coastal-population-report.pdf.

NOAA. 2020. National Centers for Environmental Information. https://www.ncdc.noaa.gov.

NRC (National Research Council). 1992. *Global Environmental Change: Understanding the Human Dimensions*. Washington, DC: The National Academies Press.

NRC. 2003. *Oil in the Sea III: Inputs, Fates, and Effects*. Washington, DC: The National Academies Press.

NRC. 2008. *Public Participation in Environmental Assessment and Decision Making*. Washington, DC: The National Academies Press.

NRC. 2010a. *Informing an Effective Response to Climate Change*. Washington, DC: The National Academies Press.

NRC. 2010b. *Advancing The Science of Climate Change*. Washington, DC: The National Academies Press.

NRC. 2010c. *Verifying Greenhouse Gas Emissions: Methods to Support International Climate Agreements*. Washington, DC: The National Academies Press.

NRC. 2011. *America's Climate Choices*. Washington, DC: The National Academies Press.

NRC. 2012. *A National Strategy for Advancing Climate Modeling*. Washington, DC: The National Academies Press.

NRC. 2015. *Climate Intervention: Reflecting Sunlight to Cool Earth*. Washington, DC: The National Academies Press.

NSB (National Science Board). 2019. *The Skilled Technical Workforce: Crafting America's Science & Engineering Enterprise*. Alexandria, VA: National Science Board.

O'Grady, C. 2020. Power to the people. *Science* 370(6516):518-521. https://doi.org/10.1126/science.370.6516.518.

O'Neill, B. C., C. Tebaldi, D. P. van Vuuren, V. Eyring, P. Fridelingstein, G. Hurtt, R. Knutti, E. Kriegler, J.-F. Lamarque, J. Lowe, J. Meehl, R. Moss, K. Riahi, and B. M. Sanderson. 2016. The Scenario Model Intercomparison Project (ScenarioMIP) for CMIP6. *Geoscientific Model Development Discussions*. https://doi.org/10.5194/gmd-2016-84.

O'Neill, B. C., T. R. Carter, K. Ebi, P. A. Harrison, E. Kemp-Benedict, K. Kok, E. Kriegler, B. L. Preston, K. Riahi, J. Sillmann, B. J. van Ruijven, D. van Vuuren, D. Carlisle, C. Conde, J. Fuglestvedt, C. Green, T. Hasegawa, J. Leininger, S. Monteith, and R. Pichs-Madruga. 2020. Achievements and needs for the climate change scenario framework. *Nature Climate Change* 10(12):1074-1084. https://doi.org/10.1038/s41558-020-00952-0.

Otto, I. M., J. F. Donges, R. Cremades, A. Bhowmik, R. J. Hewitt, W. Lucht, J. Rockström, F. Allerberger, M. McCaffrey, S. S. P. Doe, A. Lenferna, N. Morán, D. P. van Vuuren, and H. J. Schellnhuber. 2020. Social tipping dynamics for stabilizing Earth's climate by 2050. *Proceedings of the National Academy of Sciences* 117(5):2354-2365. https://doi.org/10.1073/pnas.1900577117.

Pauli, B. J. 2020. The Flint water crisis. *WIREs Water* 7(3):e1420. https://doi.org/10.1002/wat2.1420.

Pearson, A. R., J. P. Schuldt, R. Romero-Canyas, M. T. Ballew, and D. Larson-Konar. 2018. Diverse segments of the US public underestimate the environmental concerns of minority and low-income Americans. *Proceedings of the National Academy of Sciences* 115(49):12429-12434. https://doi.org/10.1073/pnas.1804698115.

Popp, A. L., S. R. Lutz, S. Khatami, T. H. M. van Emmerik, and W. J. M. Knoben. 2019. A global survey on the perceptions and impacts of gender inequality in the earth and space sciences. *Earth and Space Science* 6(8):1460-1468. https://doi.org/10.1029/2019EA000706.

Prein, A. F., W. Langhans, G. Fosser, A. Ferrone, N. Ban, K. Goergen, M. Keller, M. Tölle, O. Gutjahr, F. Feser, E. Brisson, S. Kollet, J. Schmidli, N. P. M. van Lipzig, and R. Leung. 2015. A review on regional convection-permitting climate modeling: Demonstrations, prospects, and challenges. *Reviews of Geophysics* 53(2):323-361. https://doi.org/10.1002/2014RG000475.

PWC (PricewaterhouseCoopers). 2015. *The World in 2050: Will the Shift in Global Economic Power Continue?* London: PricewaterhouseCoopers LLP.

Ranger, N., and F. Niehörster. 2012. Deep uncertainty in long-term hurricane risk: Scenario generation and implications for future climate experiments. *Global Environmental Change* 22(3):703-712. https://doi.org/10.1016/j.gloenvcha.2012.03.009.

Reed, D. A., and J. Dongarra. 2015. Exascale computing and big data. *Communications of the ACM* 58(7):56–68. https://doi.org/10.1145/2699414.

Reichstein, M., G. Camps-Valls, B. Stevens, M. Jung, J. Denzler, N. Carvalhais, and Prabhat. 2019. Deep learning and process understanding for data-driven Earth system science. *Nature* 566(7743):195-204. https://doi.org/10.1038/s41586-019-0912-1.

Roelich, K., and J. Giesekam. 2019. Decision making under uncertainty in climate change mitigation: Introducing multiple actor motivations, agency and influence. *Climate Policy* 19(2):175-188. https://doi.org/10.1080/14693062.2018.1479238.

Rushing, C. S., M. Rubenstein, J. E. Lyons, and M. C. Runge. 2020. Using value of information to prioritize research needs for migratory bird management under climate change: A case study using federal land acquisition in the United States. *Biological Reviews* 95(4):1109-1130. https://doi.org/https://doi.org/10.1111/brv.12602.

Sandifer, P. A., and G. I. Scott. 2021. Coastlines, coastal cities, and climate change: A perspective on urgent research needs in the United States. *Frontiers in Marine Science* 8(631986). https://doi.org/10.3389/fmars.2921.631986.

Sandifer, P. A., A. H. Walker, M. Finucane, H. M. Solo-Gabriele, A. E. Ferguson, M. Partyka, K. Wowk, R. Caffey, and D. Yoskowitz. 2020. Human health and socioeconomic effects of the Deepwater Horizon oil spill in the Gulf of Mexico. *Oceanography* (in review).

Schlosberg, D., and L. B. Collins. 2014. From environmental to climate justice: Climate change and the discourse of environmental justice. *WIREs Climate Change* 5(3):359-374. https://doi.org/https://doi.org/10.1002/wcc.275.

Schweizer, V. J., and B. C. O'Neill. 2014. Systematic construction of global socioeconomic pathways using internally consistent element combinations. *Climatic Change* 122(3):431-445. https://doi.org/10.1007/s10584-013-0908-z.

Semenza, J. C., J. Rocklöv, P. Penttinen, and E. Lindgren. 2016. Observed and projected drivers of emerging infectious diseases in Europe. *Annals of the New York Academy of Sciences* 1382(1):73-83. https://doi.org/10.1111/nyas.13132.

Shepherd, T. G., E. Boyd, R. A. Calel, S. C. Chapman, S. Dessai, I. M. Dima-West, H. J. Fowler, R. James, D. Maraun, O. Martius, C. A. Senior, A. H. Sobel, D. A. Stainforth, S. F. B. Tett, K. E. Trenberth, B. J. J. M. van den Hurk, N. W. Watkins, R. L. Wilby, and D. A. Zenghelis. 2018. Storylines: an alternative approach to representing uncertainty in physical aspects of climate change. *Climatic Change* 151(3):555-571. https://doi.org/10.1007/s10584-018-2317-9.

Shwom, R. 2020. *Social Tipping Points in the Climate System: Theorizing Conditions and Mechanisms of Rapid Social Change.* New Brunswick, NJ: Rutgers Energy Institute.

Siders, A. R. 2019. Managed retreat in the United States. *One Earth* 1(2):216-225. https://doi.org/10.1016/j.oneear.2019.09.008.

Siders, A. R., and J. M. Keenan. 2020. Variables shaping coastal adaptation decisions to armor, nourish, and retreat in North Carolina. *Ocean & Coastal Management* 183:105023. https://doi.org/https://doi.org/10.1016/j.ocecoaman.2019.105023.

Siders, A. R., M. Hino, and K. J. Mach. 2019. The case for strategic and managed climate retreat. *Science* 365(6455):761-763. https://doi.org/10.1126/science.aax8346.

Sinay, L., and R. W. B. Carter. 2020. Climate change adaptation options for coastal communities and local governments. *Climate* 8(1):7. https://doi.org/https://doi.org/10.3390/cli8010007.

Singu, S., A. Acharya, K. Challagundla, and S. N. Byrareddy. 2020. Impact of social determinants of health on the emerging COVID-19 pandemic in the United States. *Frontiers in Public Health* 8(406). https://doi.org/10.3389/fpubh.2020.00406.

Smith, S. R., I. Christie, and R. Willis. 2020. Social tipping intervention strategies for rapid decarbonization need to consider how change happens. *Proceedings of the National Academy of Sciences* 117(20):10629-10630. https://doi.org/10.1073/pnas.2002331117.

Steffen, W., K. Richardson, J. Rockström, H. J. Schellnhuber, O. P. Dube, S. Dutreuil, T. M. Lenton, and J. Lubchenco. 2020. The emergence and evolution of Earth System Science. *Nature Reviews Earth & Environment* 1(1):54-63. https://doi.org/10.1038/s43017-019-0005-6.

Sterling, E. J., M. Zellner, K. E. Jenni, K. Leong, P. D. Glynn, T. K. BenDor, P. Bommel, K. Hubacek, A. J. Jetter, R. Jordan, L. S. Olabisi, M. Paolisso, and S. Gray. 2019. Try, try again: Lessons learned from success and failure in participatory modeling. *Elementa, Science of the Anthropocene* 7(1):9. https://doi.org/10.1525/elementa.347.

Stern, P. C., and T. Dietz. 2020. A broader social science research agenda on sustainability: Nongovernmental influences on climate footprints. *Energy Research & Social Science* 60:101401. https://doi.org/https://doi.org/10.1016/j.erss.2019.101401.

Stern, P. C., K. L. Ebi, R. Leichenko, R. S. Olson, J. D. Steinbruner, and R. Lempert. 2013. Managing risk with climate vulnerability science. *Nature Climate Change* 3(7):607-609. https://doi.org/10.1038/nclimate1929.

Stokes, D. E. 1997. *Pasteur's Quadrant: Basic Science and Technological Innovation.* Washington, DC: Brookings Institution Press.

Swain, D. L., O. E. J. Wing, P. D. Bates, J. M. Done, K. A. Johnson, and D. R. Cameron. 2020. Increased Flood Exposure Due to Climate Change and Population Growth in the United States. *Earth's Future* 8(11):e2020EF001778. https://doi.org/https://doi.org/10.1029/2020EF001778.

Sweet, W., G. Dusek, D. March, G. Carbin, and J. Marra. 2019. *2018 State of High Tide Flooding with a 2019 Outlook*. NOAA Technical Report NOS CO-OPS 090. Silver Spring, MD: NOAA National Ocean Service, Center for Operational Oceanographic Products and Services. https://tidesandcurrents.noaa.gov/publications/Techrpt_090_2018_State_of_US_HighTide-Flooding_with_a_2019_Outlook_Final.pdf.

Tang, C. Q., Y.-F. Dong, S. Herrando-Moraira, T. Matsui, H. Ohashi, L.-Y. He, K. Nakao, N. Tanaka, M. Tomita, X.-S. Li, H.-Z. Yan, M.-C. Peng, J. Hu, R.-H. Yang, W.-J. Li, K. Yan, X. Hou, Z.-Y. Zhang, and J. López-Pujol. 2017. Potential effects of climate change on geographic distribution of the Tertiary relict tree species Davidia involucrata in China. *Scientific Reports* 7(1):43822. https://doi.org/10.1038/srep43822.

Thomas, T. K., J. Bell, D. Bruden, M. Hawley, and M. Brubaker. 2013. Washeteria closures, infectious disease and community health in rural Alaska: A review of clinical data in Kivalina, Alaska. *International Journal of Circumpolar Health* 72(1):21233. https://doi.org/10.3402/ijch.v72i0.21233.

Thompson, J. R., J. S. Plisinski, K. F. Lambert, M. J. Duveneck, L. Morreale, M. McBride, M. G. MacLean, M. Weiss, and L. Lee. 2020. Spatial simulation of codesigned land cover change scenarios in New England: Alternative futures and their consequences for conservation priorities. *Earth's Future* 8(7):e2019EF001348. https://doi.org/10.1029/2019EF001348.

Tosun, J., and J. J. Schoenefeld. 2017. Collective climate action and networked climate governance. *WIREs Climate Change* 8(1):e440. https://doi.org/10.1002/wcc.440.

Tucker, D., D. K. Ginther, and J. A. Winkler. 2009. Gender issues among academic AMS members: Comparisons with the 1993 Membership Survey. *Bulletin of the American Meteorological Society* 90(8):1180-1191. https://doi.org/10.1175/2009BAMS2538.1.

UCS (Union of Concerned Scientists). 2014. Causes of Drought: What's the Climate Connection? Reports & Media / Explainer, April 10. https://www.ucsusa.org/resources/drought-and-climate-change.

UCS. 2016. The US Military on the Front Lines of Rising Seas: Executive Summary. https://www.ucsusa.org/sites/default/files/attach/2016/07/us-military-on-front-lines-of-rising-seas_all-materials.pdf.

UN (United Nations). 2019. World Population Prospects 2019. https://population.un.org/wpp.

UN Interagency Framework Team for Preventive Action. 2012. Toolkit and Guidance for Preventing and Managing Land and Natural Resources Conflict: Renewable Resources and Conflict. https://www.un.org/en/land-natural-resources-conflict/pdfs/GN_Renew.pdf.

UN LEG. 2019. *Open NAPs*. UN Framework Convention on Climate Change, LDC Expert Group, Policy Brief No. 1, May. https://unfccc.int/sites/default/files/resource/opennapbrief.pdf.

UN Trust Fund for Human Security. 2016. *Human Security Handbook: An integrated approach for the realization of the Sustainable Development Goals and the priority areas of the international community and the United Nations system*. New York: United Nations.

UNDRR (United Nations Office for Disaster Risk Reduction). 2021. What is the Sendai Framework for Disaster Risk Reduction? https://www.undrr.org/implementing-sendai-framework/what-sendai-framework.

UNEP (United Nations Environment Programme). 2013. Climate Change, Water Shortages, Biodiversity Loss, Will Have Growing Impacts on Global Business. UN Report, Press Release. https://www.unenvironment.org/news-and-stories/press-release/climate-change-water-shortages-biodiversity-loss-will-have-growing.

USGCRP (US Global Change Research Program). 2012. *The National Global Change Research Plan 2012-2021: A Strategic Plan for the U.S. Global Change Research Program*. Washington, DC: US Global Change Research Program.

USGCRP. 2016. *The Impacts of Climate Change on Human Health in the United States: A Scientific Assessment*. Crimmins, A., J. Balbus, J.L. Gamble, C.B. Beard, J.E. Bell, D. Dodgen, R.J. Eisen, N. Fann, M.D. Hawkins, S.C. Herring, L. Jantarasami, D.M. Mills, S. Saha, M.C. Sarofim, J. Trtanj, and L. Ziska, eds. Washington, DC: U.S. Global Change Research Program.

USGCRP. 2017. *Climate Science Special Report: Fourth National Climate Assessment, Volume I*. D. J. Wuebbles, D. W. Fahey, K. A. Hibbard, D. J. Dokken, B. C. Stewart, and T. K. Maycock, eds. Washington, DC: US Global Change Research Program.

USGCRP. 2018. *Impacts, Risks, and Adaptation in the United States: Fourth National Climate Assessment, Volume II*. Washington, DC: US Global Change Research Program.

USGCRP. 2020. *Our Changing Planet: The U.S. Global Change Research Program for Fiscal Year 2020*. Washington, DC: US Global Change Research Program.

van Ginkel, K. C. H., W. J. W. Botzen, M. Haasnoot, G. Bachner, K. W. Steininger, J. Hinkel, P. Watkiss, E. Boere, A. Jeuken, E. S. de Murieta, and F. Bosello. 2020. Climate change induced socio-economic tipping points: Review and stakeholder consultation for policy relevant research. *Environmental Research Letters* 15(2):023001. https://doi.org/10.1088/1748-9326/ab6395.

Vandenbergh, M. P., and J. M. Gilligan. 2017. *Beyond Politics: The Private Governance Response to Climate Change*. Cambridge, UK: Cambridge University Press.

Wara, M. 2015. Instrument choice, carbon emissions, and information. *Michigan Journal of Environmental and Administrative Law* 4(2).

Wara, M., D. Cullenward, and R. Teitelbaum. 2015. Peak electricity and the clean power plan. *The Electricity Journal* 28(4):18-27. https://doi.org/10.1016/j.tej.2015.04.006.

Werrell, C., and F. Femia. 2019. *The Responsibility to Prepare and Prevent: A Climate Security Governance Framework for the 21st Century*. Washington, DC: Center for Climate and Security. https://climateandsecurity.org/wp-content/uploads/2019/10/the-responsibility-to-prepare-and-prevent_a-climate-security-governance-framework-for-the-21st-century_2019_10.pdf.

Weyant, J. 2017. Some contributions of integrated assessment models of global climate change. *Review of Environmental Economics and Policy* 11(1):115-137. https://doi.org/10.1093/reep/rew018.

WHO (World Health Organization). 2003. Climate change and infectious diseases. In *Climate Change and Human Health—Risks and Responses*. A. J. McMichael, ed. Geneva: World Health Organization. https://www.who.int/globalchange/climate/en/chapter6.pdf.

World Economic Forum. 2020. *The Global Risks Report 2020*. 14th Edition. Geneva, Switzerland: World Economic Forum. https://www.weforum.org/reports/the-global-risks-report-2020.

Yang, Q., T. H. Dixon, P. G. Myers, J. Bonin, D. Chambers, M. R. van den Broeke, M. H. Ribergaard, and J. Mortensen. 2016. Recent increases in Arctic freshwater flux affects Labrador Sea convection and Atlantic overturning circulation. *Nature Communications* 7(1):10525. https://doi.org/10.1038/ncomms10525.

Zhong, H., Z. Tan, Y. He, L. Xie, and C. Kang. 2020. Implications of COVID-19 for the electricity industry: A comprehensive review. *CSEE Journal of Power and Energy Systems* 6(3):489-495. https://doi.org/10.17775/CSEEJPES.2020.02500.

Zhu, C., K. Kobayashi, I. Loladze, J. Zhu, Q. Jiang, X. Xu, G. Liu, S. Seneweera, K. L. Ebi, A. Drewnowski, N. K. Fukagawa, and L. H. Ziska. 2018. Carbon dioxide (CO_2) levels this century will alter the protein, micronutrients, and vitamin content of rice grains with potential health consequences for the poorest rice-dependent countries. *Science Advances* 4(5):eaaq1012. https://doi.org/10.1126/sciadv.aaq1012.

Ziegler, T. B., C. M. Coombe, Z. E. Rowe, S. J. Clark, C. J. Gronlund, M. Lee, A. Palacios, L. S. Larsen, T. G. Reames, J. Schott, G. O. Williams, and M. S. O'Neill. 2019. Shifting from "community-placed" to "community-based" research to advance health equity: A case study of the Heatwaves, Housing, and Health: Increasing Climate Resiliency in Detroit (HHH) partnership. *International Journal of Environmental Research and Public Health* 16(18). https://doi.org/10.3390/ijerph16183310.

APPENDIX A

Statement of Task

The Advisory Committee will author a short report that provides high-level guidance on research priorities for the U.S. Global Change Research Program (USGCRP) as a way to proactively meet their charge to advise the Program on broad, Program-wide issues, and to identify topics of importance to the global change science community. The Committee will consider how USGCRP can best meet the mandate of the US Global Change Research Act of 1990, given the significant climate change impacts happening today and projected to increase in the future. In addressing its charge, the Committee will:

- Provide a short update on accomplishments of USGCRP to date;

- Identify the most critical global change risks and uncertainties facing the nation and the world in the next 5-10 years;

- Recommend priorities for research needed to advance understanding of these risks and uncertainties and to support decision making at local to national scales; and

- Discuss opportunities for USGCRP participating agencies and other partners to advance the identified research priorities and applications to decision contexts, including new approaches for better linking the process of scientific deliberations with people who use information.

APPENDIX B

Committee Member Biographies

Jerry M. Melillo (*Chair, NAS*) is a Distinguished Scientist at the Marine Biological Laboratory whose work focuses on understanding the impacts of human activities on the biogeochemistry of ecological systems using a combination of field studies and simulation modeling. His field studies include soil warming experiments at the Harvard Forest in central Massachusetts. Dr. Melillo and his team have developed and used a simulation model called the Terrestrial Ecosystem Model (TEM) to consider the impacts of various aspects of global change on the structure and function of terrestrial ecosystems across the globe. TEM is part of the Integrated Global Systems Model, an integrated assessment model, based at the Massachusetts Institute of Technology.

Kristie L. Ebi (*Vice Chair*) is a Professor in the Department of Global Health and in the Department of Environmental and Occupational Health Sciences, University of Washington. She has been conducting research and practice on the health risks of climate variability and change for nearly 25 years, focusing on understanding sources of vulnerability; estimating current and future health risks of climate change; designing adaptation policies and measures to reduce risks in multistressor environments; and estimating the health co-benefits of mitigation policies. She has supported multiple countries in Central America, Europe, Africa, Asia, and the Pacific in assessing their vulnerabilities and implementing adaptation policies and programs. She has been an author on multiple national and international climate change assessments, including the Fourth U.S. National Climate Assessment and the Intergovernmental Panel on Climate Change Special Report on Global Warming of 1.5°C. She is co-chair of the National Academies of Sciences, Engineering, and Medicine's Board on Environment and Society, the International Committee on New Integrated Climate Change Assessment Scenarios, and the Future Earth Health Knowledge Action Network. She is a member of the Earth Commission and of the Earth League. Dr. Ebi's scientific training includes an M.S. in toxicology and a Ph.D. and a Master's of Public Health in epidemiology, and postgraduate research at the London School of Hygiene and Tropical Medicine. She edited fours books on aspects of climate change and published more than 200 papers.

Arrietta Chakos is a public policy advisor on urban resilience. She works on community resilience strategies and multisectoral engagement. Her work with San Francisco, Palo Alto, and regional institutions, such as the Association of Bay Area Governments, focuses on disaster readiness and community resilience. She was an appointed member of the Resilience Roundtable at the National Academy of Sciences and chaired the Housner Fellow committee at the Earthquake Engineering Research Institute. Ms. Chakos served as research director of the Harvard Kennedy School's Acting in Time Advance Recovery Project. She was assistant city manager in Berkeley, California, directing innovative risk mitigation initiatives, intergovernmental coordination, and multi-institutional negotiations. Specialties include urban resilience strategies, public policy development, climate change adaptation, disaster risk assessment and loss estimates, mitigation and risk financing, strategic fiscal planning, multiparty negotiations, and municipal government operations.

Peter Daszak (NAM) is President of EcoHealth Alliance (EHA), a U.S.-based organization that conducts research and outreach programs on global health, conservation, and international development. Dr. Daszak's research has been instrumental in identifying and predicting the impact of emerging diseases across the globe. Dr. Daszak is Chair of the National Academies of Sciences, Engineering, and Medicine's Forum on Microbial Threats. He is a member of the National Academies' Advisory Committee to the U.S. Global Change Research Program, the Supervisory Board of the One Health Platform, the One Health Commission Council of Advisors, the Center of Excellence for Emerging Zoonotic Animal Diseases External Advisory Board, the Cosmos Club, and the Advisory Council of the Bridge Collaborative. He has served on the Institute of Medicine Committee on Global Surveillance for Emerging Zoonoses, the National Research Council Committee on the Future of Veterinary Research, and the International Standing Advisory Board of the Australian Biosecurity Cooperative Research Centre; and he has advised the Director for Medical Preparedness Policy on the White House National Security Staff on global health issues. Dr. Daszak is a regular advisor to the World Health Organization on pathogen prioritization for research and development. Dr. Daszak won the 2000 Commonwealth Scientific and Industrial Research Organisation medal for collaborative research on the discovery of amphibian chytridiomycosis; is the EHA institutional lead for the U.S. Agency for International Development-Emerging Pandemic Threats-PREDICT; is on the Editorial Boards of *Conservation Biology*, *One Health*, and *Transactions of the Royal Society of Tropical Medicine & Hygiene*; and is Editor-in-Chief of the journal *EcoHealth*. He has authored more than 300 scientific papers, and his work has been the focus of extensive media coverage.

Thomas Dietz is University Distinguished Professor of Sociology, Environmental Science and Policy and Animal Studes at Michigan State University. He holds a Ph.D. in ecology from the University of California at Davis. He is a Fellow of the American Association for the Advancement of Science, and has been awarded the Sustainability Science Award of the Ecological Society of America, the Distinguished Contribution Award and the Outstanding Publication Award from the American Sociological Association Section on Environment, Technology and Society. He chaired the National Research Council (NRC) Committee on Human Dimensions of Global Change and the panel on Public Participation in Environmental Assessment and Decision Making and served as Vice Chair of the NRC Panel on Advancing the Science of Climate Change (America's Climate Choices). His current research examines the human driving forces of environmental change, environmental values, and the interplay between science and democracy in environmental issues.

Philip B. Duffy is President and Executive Director of Woodwell Climate Research Center (formerly Woods Hole Research Center). Prior to joining Woodwell, Dr. Duffy served as a Senior Policy Analyst in the White House Office of Science and Technology Policy and as a Senior Advisor in the White House National Science and Technology Council. In these roles he was involved in international climate negotiations, domestic and international climate policy, and coordination of U.S. global change research. Before joining the White House, Dr. Duffy was Chief Scientist at Climate Central, an organization dedicated to increasing public understanding and awareness of climate change. Dr. Duffy has held senior research positions with the Lawrence Livermore National Laboratory, and visiting positions at the Carnegie Institution for Science and the Woods Institute for the Environment at Stanford University. He has a bachelor's degree *magna cum laude* from Harvard in astrophysics and a Ph.D. in applied physics from Stanford University.

Baruch Fischhoff (NAS, NAM) is Howard Heinz University Professor, Department of Engineering and Public Policy and Institute for Politics and Strategy, Carnegie Mellon University (CMU). A graduate of the Detroit Public Schools, he holds a B.S. (mathematics, psychology) from Wayne State University and a Ph.D. (psychology) from the Hebrew University of Jerusalem. He is a member of the National Academy of Sciences and of the National Academy of Medicine. He is past President of the Society for Judgment and Decision Making and of the Society for Risk Analysis. He has chaired the Food and Drug Administration Risk Communication Advisory Committee and been a member of the Eugene (Oregon) Commission on the Rights of Women, the U.S. Department of Homeland Security Science and Technology Advisory Committee, and the U.S. Environmental Protection Agency Scientific Advisory Board, where he chaired the Homeland Security Advisory Committee. He has received the American Psychological Association (APA) Award for Distinguished Contribution to Psychology, CMU's Ryan Award for Teaching, an honorary Doctorate of Humanities from Lund University, and an Andrew Carnegie Fellowship. He is a Fellow of APA, the Association for Psychological Science, Society of Experimental Psychologists, and Society for Risk Analysis. His books include *Acceptable Risk*, *Risk: A Very Short Introduction*, *Judgment and Decision Making*, *A Two-State Solution in the Middle East*, *Counting Civilian Casualties*, and *Com-*

municating Risks and Benefits. He has co-chaired three National Academy Colloquia on the Science of Science Communication.

Paul Fleming leads the Global Water Program for Microsoft. Paul joined Microsoft to build its corporate water stewardship program and has helped establish Microsoft as a leader in the corporate water stewardship space. In addition to driving the company's operational water commitments, Mr. Fleming drives collaborative partnerships with other companies and nongovernmental organizations and serves as the company's water subject matter expert, advising business groups on water issues. He is on the leadership committee of the Water Resilience Coalition, a group of 18 companies focused on collective action to improve conditions in water-stressed regions around the world, and serves on the steering committee of the CEO Water Mandate. Previously, Mr. Fleming developed and directed the Seattle Public Utilities' (SPU's) Climate Resiliency Group, where he was responsible for directing SPU's climate research initiatives, assessing climate risks, mainstreaming adaptation and mitigation strategies, and establishing collaborative partnerships. Mr. Fleming has been an active participant in several national and international efforts focused on water and climate change. He contributed to the 2014 U.S. National Climate Assessment, serving as a Convening Lead Author of the Water Resources chapter and the Sustained Assessment Special Report and a Lead Author of the Adaptation chapter. He is a Past Chair of the Water Utility Climate Alliance and chaired the Project Advisory Board of a research project focused on climate change and water management funded through the EU Horizon 2020 Program. Mr. Fleming has a B.A. in economics from Duke University and an M.B.A. from the University of Washington.

Sherri W. Goodman is an executive, lawyer, former defense official, and Senate Armed Services Committee staff professional. She is currently a Senior Fellow at the Woodrow Wilson International Center for Scholars and at CNA. Most recently, she served as the President and CEO of the Consortium for Ocean Leadership, which manages federally funded science and technology programs and whose members are the nation's leading ocean science research institutions. Ms. Goodman previously served as Senior Vice President, General Counsel and Corporate Secretary of CNA, a research organization for national security leaders and public sector organizations. She is the founder and Executive Director of the CNA Military Advisory Board, whose landmark reports include National Security and the Threat of Climate Change (2007), Powering America's Economy: Energy Innovation at the Crossroads of National Security Challenges (2010), National Security and the Accelerating Risks of Climate Change (2014), and Advanced Energy and US National Security (2017), among others. Ms. Goodman served as the first Deputy Undersecretary of Defense (Environmental Security), responsible for global environmental, energy efficiency, safety, and occupational health programs and policies of the U.S. Department of Defense. She served on the staff of the Senate Armed Services Committee, where she was responsible for oversight of the U.S. Department of Energy's nuclear weapons complex, including the national laboratories and the defense environmental management program. Ms. Goodman is a member of the Secretary of State's International Security Advisory Board, for which she co-chaired a report on Arctic Security. She serves on the boards of the Atlantic Council, the Adrienne Arsht Resilience Center, the Center for Climate and Security, the Joint Ocean Leadership Initiative, the Marshall Legacy Institute, the University Cooperation for Atmospheric Research, and the U.S. Water Partnership. She is a life member of the Council on Foreign Relations (CFR) and a member of the CFR Arctic Task Force. She has served as a Trustee of the Woods Hole Oceanographic Institution and on the Committee on Conscience of the U.S. Holocaust Memorial Museum. Ms. Goodman is a graduate of Amherst College; she holds degrees from Harvard Law School and Harvard's Kennedy School of Government.

Nancy B. Grimm (NAS) is Regents Professor and Virginia M. Ullman Professor of Ecology at Arizona State University. She studies the interaction of climate variation and change, human activities, and ecosystems. Her long-term research focuses on how disturbances (e.g., flooding or drying) affect the structure and processes of desert streams, how chemical elements move through and cycle within both desert streams and cities, and how storm water infrastructure affects water and material movement across an urban landscape. Dr. Grimm was the founding director of the Central Arizona-Phoenix Long-Term Ecological Research program—an interdisciplinary study by ecologists, engineers, physical and social scientists—and currently co-directs the Urban Resilience to Extremes Sustainability Research Network. In the latter capacity, she works to help cities develop future visions and strategies to increase resilience in the face of extreme events. She is a fellow of the American Association for the Advancement of Science, the American Geophysical Union, Ecological Society of America (ESA), and the Society for Freshwater Science (SFS). She is past president of the ESA and the SFS, and was an author on the second and third National Climate Assessments. She is a graduate of Hampshire College, and received her Ph.D. in 1985 from Arizona State University.

Henry D. Jacoby is Professor of Management, emeritus, in the Massachusetts Institute of Technology (MIT) Sloan School of Management and former Co-Director of the MIT Joint Program on the Science and Policy of Global Change, which is focused on the integration of the natural and social sciences and policy analysis in application to the threat of global climate change. An undergraduate mechanical engineer at the University of Texas at Austin, he holds a Ph.D. in economics from Harvard University and Doctorats Honoris Causa from the University of Geneva. At Harvard he served on the faculties of the Department of Economics and the Kennedy School of Government and as Director of the Environmental Systems Program. At MIT he has been Director of the Center for Energy and Environmental Policy Research, Associate Director of the Energy Laboratory, and Chair of the Faculty. Professional activities have included the U.S. National Petroleum Council, the Nuclear Fuels Working Group of the Atlantic Council, and the Scientific Committee of the International Geosphere-Biosphere Program.

Linda O. Mearns is Head of the Regional Integrated Sciences Collective within the Computational and Information Systems Lab and the Research Applications Lab, and Senior Scientist at the National Center for Atmospheric Research, Boulder, Colorado. She served as Director of the Institute for the Study of Society and Environment for 3 years ending in 2008. She holds a Ph.D. in geography/climatology from the University of California, Los Angeles. She has performed research and published mainly in the areas of climate change scenario formation, quantifying uncertainties, and climate change impacts on agro-ecosystems. She has particularly worked extensively with regional climate models. She has been an author in the Intergovernmental Panel on Climate Change's Climate Change 1995, 2001, 2007, 2014, and current (2021) Assessments regarding climate variability, impacts of climate change on agriculture, regional projections of climate change, climate scenarios, and uncertainty in future projections of climate change. For the Sixth Assessment Report, she is a lead author of the Atlas in Working Group I and a Review Editor for the North America Chapter in Working Group II. She led the multiagency-supported North American Regional Climate Change Assessment Program, which provided multiple high-resolution climate change scenarios for the North American impacts community and is currently the co-Chair of the NA-CORDEX regional modeling program. She has been a member of the National Research Council Climate Research Committee, the National Academy of Sciences (NAS) Panel on Adaptation of the America's Climate Choices Program, and the NAS Human Dimensions of Global Change Committee. She has worked extensively with resource managers (e.g., water resource managers and ecologists) to form climate change scenarios for use in adaptation planning.

Richard H. Moss is a senior scientist at Pacific Northwest National Laboratory's Joint Global Change Research Institute, and holds visiting/adjunct appointments at Princeton University and the University of Maryland. Dr. Moss's research focuses on (1) vulnerability assessment and adaptation to global change, (2) uncertainty characterization and communication, and (3) scenarios. His current research on global change impacts focuses on multisector/multiscale modeling of global change impacts and responses. Previously he served as Director of the U.S. Global Change Research Program (spanning the Clinton and G.W. Bush Administrations), head of technical support for Working Group II of the Intergovernmental Panel on Climate Change, and director of climate/energy at the United Nations Foundation and the World Wildlife Fund (United States). He received his Ph.D. from Princeton University in public and international affairs.

Margo Oge is an author and former director of the Office of Transportation and Air Quality of the U.S. Environmental Protection Agency (EPA). In her new book *Driving the Future: Combating Climate Change with Cleaner, Smarter Cars,* Ms. Oge chronicles the political and regulatory history that led to America's first formal climate action using regulation to reduce emissions through innovation in car design and portrays a future where clean, intelligent vehicles with lighter frames and alternative power trains will radically reduce carbon pollution. Ms. Oge retired as director of the Office of Transportation and Air Quality after 32 years with EPA. While at EPA, she was a chief architect of the most important improvements of air quality from the transportation sector ever, resulting in the prevention of 40,000 premature deaths and hundreds of thousands of cases of respiratory illness. She led EPA's first-ever national greenhouse gas (GHG) emission standards for cars and heavy-duty trucks to double fuel efficiency by 2025 and reduce GHG emissions by 50 percent. She received Presidential Awards from Presidents Bill Clinton and George W. Bush and numerous environmental and industry awards. In commending her achievements, President Obama wrote, "Under your tireless leadership, we have realized significant environmental achievements in the transportation sector, from making diesel fuels cleaner to finalizing the most aggressive fuel economy standards for cars and trucks out to the model year 2025." Ms. Oge serves as a Distinguished Fellow with ClimateWorks, a nongovernmental organization that works globally to strengthen philanthropy's response to climate change. She serves on the International Sustainability Council of the Volkswagen and is the Vice Chairman of the Board of DeltaWing Technologies, which is creating a new, high-efficiency passenger car based on the DeltaWing race car. She is a member of the board of the Union of Concerned Scientists, the International Council on Clean Transportation, and the Alliance of Climate Education. She serves on the National Academies of Sciences, Engineering, and Medicine's Board on Energy and Environmental Systems and the Advisory Committee on Climate Change Research as well as the U.S. Department of Energy Advisory Committee on Hydrogen and Fuel Cells. She is also an advisor to Square Roots, a life science company. Ms. Oge has an M.S. in engineering from the University of Massachusetts–Lowell and attended George Washington and Harvard Universities.

S. George H. Philander (NAS) is Knox Taylor Professor of Geosciences at Princeton University. He received his bachelor's degree from the University of Cape Town in 1962 and his Ph.D. in mathematics from Harvard University in 1970 with a thesis titled "The Equatorial Dynamics of a Homogeneous Ocean." After completing 1 year as a fellow at the Massachusetts Institute of Technology, he spent 6 years as a research associate in the Geophysics Fluid Dynamics Program at Princeton University where in 1990 he became a professor in the Department of Geosciences. Dr. Philander has been a visiting professor at the Museum National d'Histoire Naturelle in Paris, a distinguished visiting scientist at the Jet Propulsion Laboratory and the California Institute of Technology, a consultant to the World Meteorological Organization in Switzerland, and a trustee of the University Corporation for Atmospheric Research in Boulder, Colorado. At Princeton, Dr. Philander became chairman of his department in 1994 and presently serves as director of its Atmospheric and Oceanic Sciences Program. He has been a fellow of the American Meteorological Society, the American Geological Union, and, in 2003, he was elected to the American Academy of Arts and Sciences. Among his publications are his books *El Niño, La Niña, and the Southern Oscillation* (San Diego: Academic Press, 1990); *Is the Temperature Rising?: The Uncertain Science of Global Warming* (Princeton, N.J.: Princeton University Press, 1998); and *Our Affair With El Niño: How We Transformed an Enchanting Peruvian Current into a Global Climate Hazard* (N.J.: Princeton University Press, 2004).

Benjamin L. Preston is a senior policy researcher at the RAND Corporation and director of RAND's Community Health and Environmental Policy Program. His recent research efforts include understanding the role of knowledge in climate risk management, evaluation of disaster recovery options and their implementation in Puerto Rico and the U.S. Virgin Islands, scenario analysis for a low-carbon future, and assessment of the environmental justice dimensions of climate risk. Previously, he held research positions with the Climate Change Science Institute at Oak Ridge National Laboratory, the Commonwealth Scientific and Industrial Research Organisation's Division of Marine and Atmospheric Research, and the Pew Center on Global Climate Change. In 2015, he received the American Geophysical Union's Falkenberg Award, and from 2016 to 2017 he was one of the American Association for the Advancement of Science's inaugural Leshner Leadership Fellows. Dr. Preston has contributed to national and international scientific assessments including the U.S. National Climate Assessment, the Intergovernmental Panel on Climate Change's Fifth and Sixth Assessment Reports, the U.S. Global Change Research Program's second State of the Carbon Cycle Report, and the Arctic Monitoring and Assessment Program's Adaptation Actions for a Changing Arctic. He currently serves as co-editor-in-chief for the Elsevier journal *Climate Risk Management*. Dr. Preston received a B.S. in biology from the College of William & Mary and a Ph.D. in environmental biology from the Georgia Institute of Technology.

Paul A. Sandifer is Director of the Center for Coastal Environmental and Human Health at the College of Charleston and Deputy Director for the Center for Oceans and Human Health and Climate Change Interactions at the University of South Carolina.

He is experienced in ecological and aquaculture research, natural resource management, science policy, and environmental health science. Previously he worked nearly 12 years in the National Oceanic and Atmospheric Administration (NOAA), overseeing the agency's Oceans and Human Health Program and as Senior Science Advisor to the NOAA Administrator and Chief Science Advisor for the National Ocean Service. Before NOAA, Dr. Sandifer worked 31 years as a scientist and manager, including as agency Director, with the South Carolina Department of Natural Resources. He served on the U.S. Commission on Ocean Policy and is an Honorary Life Member of the World Aquaculture Society and a Fellow of the American Association for the Advancement of Science and the Ecological Society of America. He received a B.S. degree in biology from the College of Charleston and Ph.D. in marine science from the University of Virginia.

Henry G. Schwartz, Jr. (NAE) is a nationally recognized civil and environmental engineering leader who spent most of his career with Sverdrup Civil Inc. (now Jacobs Civil Inc.). In 1993, Dr. Schwartz was named president and chairman, directing the transportation, public works, and environmental activities of this international engineering firm before he retired in 2003. He has served on the advisory boards for Carnegie Mellon University, Washington University in St. Louis, and the University of Texas at Austin. He is President Emeritus of the American Society of Civil Engineers, the Water Environment Federation, and the Academy of Science of St. Louis, and the founding chairman of the Water Environment Research Foundation. Elected to the National Academy of Engineering in 1997, Dr. Schwartz has served on a number of National Research Council (NRC) study committees, including the Transportation Research Board's (TRB's) Committee for a Future Strategic Highway Research Program, and on the NRC Board on Infrastructure and the Constructed Environment. He chaired the policy study committee that produced the report, Potential Impacts of Climate Change on U.S. Transportation. A convening lead author on National Climate Assessment (NCA) 2 and NCA 3, he has authored other papers focused on adaptation to climate change. For many years, he was on the TRB Executive Committee and served as Vice Chair of TRB's Subcommittee for NRC Oversight in which capacity he was the final review authority for about 100 published transportation research reports. Dr. Schwartz earned a Ph.D. from the California Institute of Technology and Master of Science and Bachelor of Science degrees from Washington University. He is a registered professional engineer.

Kathleen Segerson is a Board of Trustees Distinguished Professor of Economics and Associate Dean of The Graduate School at the University of Connecticut. She was the Head of the Department of Economics from 2001 to 2005. Dr. Segerson specializes in environmental and natural resource economics and, in particular, the economics of environmental regulation. She is currently a member of the Board of Directors of the Beijer Institute for Ecological Economics in Stockholm and a member of the U.S. National Member Organization for the International Institute for Applied Systems Analysis in Austria. She has served on the Chartered Executive Board of the U.S. Environmental Protection Agency's Science Advisory Board and was Vice Chair of the Advisory Board's Committee on Valuing the Protection of Ecological Services and Systems. She was a member of the National Research Council's (NRC's) Board on Agriculture and Natural Resources from 2009 to 2015. She has also served on several NRC study committees: the Committee on Assessing and Valuing the Services of Aquatic and Related Terrestrial Ecosystems (2002–2004), the Committee on the Causes and Management of Coastal Eutrophication (1998–2000), the Committee on Improving Principles and Guidelines for Waste Resources Planning by the U.S. Army Corps of Engineers (2008–2010), the Committee on a Study of Food Safety and Other Consequences of Publishing Establishment-Specific Data (2011), and the Review Panel on the National Climate Assessment (2012–2013). She is a Fellow of both the Agricultural and Applied Economics Association and the Association of Environmental and Resource Economists. Dr. Segerson earned a Ph.D. from Cornell University in 1984.

Brian L. Zuckerman is a Research Staff Member at the Institute for Defense Analyses Science and Technology Policy Institute (STPI). Dr. Zuckerman's areas of emphasis at STPI are in the areas of program evaluation and scientometrics, where his work focuses on federal research and development program performance and agency-wide research portfolios. Dr. Zuckerman has also analyzed federal research and development data systems and statistical data collection programs. Before joining STPI, he was a principal at C-STPS, LLC, and at the Center for Science and Technology Policy of Abt Associates, Inc. He is a former co-chair of the Research, Technology, and Development Topical Interest Group of the American Evaluation Association. Dr. Zuckerman holds a B.A. in chemistry from Harvard College and a Ph.D. in technology, management, and policy from the Massachusetts Institute of Technology.